Soybean Physiology, Agronomy, and Utilization

Contributors

W. A. Brun
Walter R. Fehr
D. R. Hicks
D. J. Hume
Harry C. Minor
A. G. Norman
F. T. Orthoefer
J. W. Tanner
D. Keith Whigham

Soybean Physiology, Agronomy, and Utilization

Edited by

A. GEOFFREY NORMAN

The University of Michigan
Ann Arbor, Michigan

ACADEMIC PRESS
New York   San Francisco   London   1978
A Subsidiary of Harcourt Brace Jovanovich, Publishers

COPYRIGHT © 1978, BY ACADEMIC PRESS, INC.
ALL RIGHTS RESERVED.
NO PART OF THIS PUBLICATION MAY BE REPRODUCED OR
TRANSMITTED IN ANY FORM OR BY ANY MEANS, ELECTRONIC
OR MECHANICAL, INCLUDING PHOTOCOPY, RECORDING, OR ANY
INFORMATION STORAGE AND RETRIEVAL SYSTEM, WITHOUT
PERMISSION IN WRITING FROM THE PUBLISHER.

ACADEMIC PRESS, INC.
111 Fifth Avenue, New York, New York 10003

United Kingdom Edition published by
ACADEMIC PRESS, INC. (LONDON) LTD.
24/28 Oval Road, London NW1 7DX

Library of Congress Cataloging in Publication Data

Main entry under title:

Soybean biology, agronomy, and utilization.

   Includes bibliographies and index.
   1. Soybean. I. Norman, Arthur Geoffrey, Date
SB205.S7S535   635'.655   78–18399
ISBN 0–12–521160–0

PRINTED IN THE UNITED STATES OF AMERICA

Contents

LIST OF CONTRIBUTORS ix
PREFACE xi

1 Background
A. G. NORMAN

I. Soybean Production 1
II. Uses and Economics of Soybean Products 12
III. The Soybean in Physiological Research 14

2 Growth and Development
D. R. HICKS

I. Plant Development 17
II. Environmental Effects on Plant Development and Performance 30
References 41

3 Assimilation
W. A. BRUN

I. Introduction 45
II. Carbon Assimilation 46
III. Nitrogen Assimilation 62
IV. Summary 72
References 73

v

4 Agronomic Characteristics and Environmental Stress
D. Keith Whigham and Harry C. Minor

I.	Introduction	78
II.	Light	78
III.	Temperature	89
IV.	Water	97
V.	Wind	102
VI.	Pests	105
VII.	Conclusions	115
	References	116

5 Breeding
Walter R. Fehr

I.	Breeding Objectives	120
II.	Steps in Cultivar Development	127
III.	Breeding Methods	132
IV.	Breeding Operations	143
V.	Blends	152
VI.	Hybrids	153
	References	155

6 Management and Production
J. W. Tanner and D. J. Hume

I.	Introduction	158
II.	Planting	159
III.	Management during the Growing Season	185
IV.	Harvesting	209
V.	Drying and Storage	214
	References	216

7 Processing and Utilization
F. T. Orthoefer

I.	Introduction	219
II.	Composition of the Seed	221

Contents

III.	Processing	228
IV.	Utilization	236
V.	Future Projections	244
	References	246

SUBJECT INDEX 247

List of Contributors

Numbers in parentheses indicate the pages on which the authors' contributions begin.

W. A. BRUN (45), Department of Agronomy and Plant Genetics, University of Minnesota, St. Paul, Minnesota 55108

WALTER R. FEHR (119), Department of Agronomy, Iowa State University, Ames, Iowa 50011

D. R. HICKS (17), Agricultural Extension Service, University of Minnesota, St. Paul, Minnesota 55108

D. J. HUME (157), Department of Crop Sciences, University of Guelph, Guelph, Ontario, Canada N1G 2W1

HARRY C. MINOR (77), Department of Agronomy, University of Illinois, Urbana, Illinois 61801

A. G. NORMAN (1), The University of Michigan, Ann Arbor, Michigan 48109

F. T. ORTHOEFER (219), A. E. Staley Manufacturing Company, Decatur, Illinois 62525

J. W. TANNER (157), Department of Crop Science, University of Guelph, Guelph, Ontario, Canada N1G 2W1

D. KEITH WHIGHAM (77), Department of Agronomy, Iowa State University, Ames, Iowa 50011

Preface

In a relatively brief period the soybean has become a major crop plant in the United States. Based on the utilization of the bean, or products therefrom, a substantial soybean industry has also developed. Its uses, agricultural and industrial, primarily depend on the high content of both protein (ca. 40%) and oil (ca. 20%) in the bean. Soybeans are a cash crop and provide a significant part of the farm income in those eight states in the Mississippi River valley that together account for 75% of the United States production. Revenues from the export of almost half the crop as beans, meal, or oil now are a strong item in the balance-of-trade figures.

These developments have sprung from and stimulated much research on the physiology, genetics, and related characteristics of the soybean plant, on the one hand, and on agronomic aspects of its management and incorporation into prevailing farm systems, on the other. The fruits of much of this research have been rather quickly put into practice. New varieties or cultivars better adapted to the physical and agronomic environment of a designated area have been produced; some were soon replaced by even better yielding or more dependable cultivars. Concurrently there has been the development of effective measures of pest and disease control.

My associates in the preparation of this book are all involved in aspects of soybean research, improvement, and utilization programs. All are fascinated, as I have been, by the characteristics and environmental responsiveness of the soybean plant. As these become better understood and as the inheritance of the qualitative and quantitative characters controlling their expression is worked out, so can the breeder develop pure lines that can be expected to perform well in specified or designated areas.

Soybean yields are good in the corn belt region of the upper Mississippi River valley but not because the environment of that region is uniquely favorable. Initially, material that suited the cornbelt was selected from the available germplasm of Oriental origin. The earlier improvement work was concentrated on developing better adapted lines from these selections. Subsequently, following similar procedures, new cultivars adapted to other requirements have been produced so that the area of profitable production in the United States has steadily expanded and may expand further. Soybeans can be grown from the equator to latitudes of more than 50°. There is a high probability that this may become a crop far more widely planted than at present, particularly in subtropical and middle latitude regions. This will in great measure depend on the recognition of the environmental elements, positive and negative, that influence the growth and development of the plant, and of the requirements of management and use that govern the adoption of a new crop.

In this book, therefore, we have attempted to cover and treat in logical sequence those factors that contribute to the potential and versatility of this useful crop plant.

A. Geoffrey Norman

1

Background

A. G. NORMAN

I. Soybean Production	1
A. In the United States	1
B. Worldwide	10
II. Uses and Economics of Soybean Products	12
III. The Soybean in Physiological Research	14

I. SOYBEAN PRODUCTION

A. In the United States

In the Orient the soybean has a long history as a crop plant for human food and animal feed, but elsewhere there has been significant production only in this century. The establishment of the soybean as a major crop in the United States has occurred since 1940, and in some other areas, such as Brazil, even more recently. Expansion into new areas is not improbable.

There is general agreement that the soybean plant has its origins in the northeastern provinces of China and Manchuria. The soybean of commerce is *Glycine max* (L.) Merrill which taxonomists believe may have developed from *Glycine ussuriensis*, Regal and Maack, a viny annual found in northern China, Korea, Taiwan, and Japan. The genus *Glycine* is large with subgenera consisting of vining perennials found in Australia, Africa, and southwest Asia. Attempts to hybridize these with *G. max* have been unsuccessful.

As botanists obtained soybean collections from China and Manchuria in the closing decades of the nineteenth century, seeds were distributed to botanical gardens and agricultural stations in western Europe and the United States. Descriptions of the growth and productivity of the soybean appeared in various scientific journals of the time. In the United States, reports were published in a number of state agricultural experiment station bulletins around the turn of the century. Nowhere outside the Orient, however, did the plant appear to have promise as a source of human food. Because of the viny habit of much of the material available, most consideration was given to its possible use as a forage crop. Its acceptance in the United States, though slow, was largely due to the interest and efforts of C. V. Piper and W. J. Morse of the United States Department of Agriculture.

The soybean varieties initially planted in the United States were selections from a large number of introductions collected at various times in the Orient. The germplasm available to the soybean breeder in the United States has the same base. Breeding programs involving plant hybridization are comparatively recent and utilize information on the types of gene action that underlie the inheritance of plant characteristics of agronomic interest. The inconspicuous flowers of the soybean are ordinarily self-fertilized, though some natural crosses do occur.

The area planted to soybeans slowly enlarged, mainly in the Mississippi Valley states, with Illinois as the leader. By 1938 it had reached 4 million hectares (10 million acres) and a processing and marketing infrastructure was well developed. Even so, on a substantial portion of this area the soybean was still grown as a forage crop, either for hay, silage, or grazing. Not until 1941 did the portion of the planting harvested for beans exceed that for forage purposes. For various reasons, primarily economic, the planting of soybeans for forage rapidly declined and has virtually disappeared except as an occasional expedient following failure of other crops.

World War II provided a strong stimulus to the planting of soybeans. By that time varieties had been developed that were well adapted to the corn belt states of Illinois, Indiana, Iowa, and Ohio. In the war-time years, economic inducements were made to farmers to increase the soybean acreage. Concurrently management experience was accumulating. Yields generally averaged under 1330 kg/ha (20 bushels/acre). Between 1935 and 1945 United States production of soybeans was quadrupled.

Since 1945 the soybean plantings have expanded fivefold and ha yields have increased so that the 1975 production was more than eightfold that of 1945. Soybeans have become a major cash crop in United States agriculture second only to corn in return to the farmer. Data on the overall United States production of soybeans by 5-year periods since 1925 are shown in Table I.

Although Illinois and the adjacent corn belt states remain the heart of United States soybean production there has been considerable expansion

TABLE I

United States Soybean Production 1925-1976[a]

Year	Harvested area[b] hectares × 1000	Production[c] metric tons × 1000
1925	168	132
1930	435	379
1935	1180	1331
1940	1945	2124
1945	4346	5258
1950	5588	8145
1955	7535	10170
1960	9573	15108
1965	13941	23016
1970	17098	30677
1975	21694	41406
1976	20017	34423

[a] Figures converted from data in Agricultural Statistics, United States Department of Agriculture.
[b] 1 Hectare = 2.47 acres.
[c] 1 Metric ton = 36.75 bushels.

both north and south, and beyond the Mississippi valley. Figure 1 shows the location of production in the mid-1970's.

It was early recognized that flowering and maturation of the soybean was related to day length and therefore to latitude. Varieties that had been selected from the introductions from the Orient differed in flowering and maturity dates. Expansion into new areas as a profitable crop depended upon the availability of seed that would best respond to the day length and other environmental conditions. An important step in soybean improvements was the establishment of maturity groups, now 12 in number, into which varieties were placed according to relative times to maturity (see Chapter 4). Breeding programs, carried out by the United States Department of Agriculture and the State Experiment Stations have increased in sophistication and effectiveness as information has accumulated on the inheritance of plant characters. Almost 100 named varieties, now called cultivars, have been registered with the Crop Science Society of America, each with detail as to source. The commercially planted varieties in the producing areas have undergone rapid replacement as newer improved cultivars appear, incorporating desirable agronomic characteristics, such as resistance to disease, lodging, or shattering. Further replacement is likely. Continuing expansion of the area of soybean production in the United States is probable. It must be borne in mind, however, that although environmental considerations may

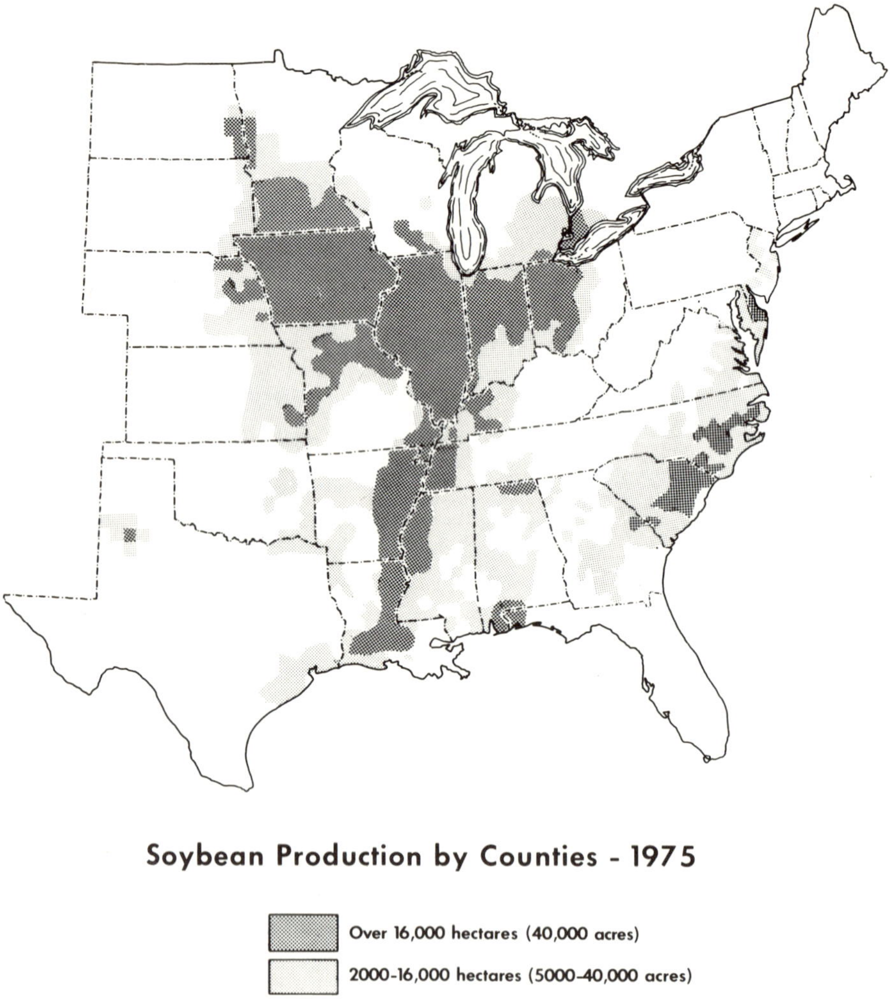

Soybean Production by Counties - 1975

Fig. 1. Location of soybean production in the United States for 1975. Prepared from county yield data. Statistical Reporting Service, United States Dept. of Agriculture.

determine where soybeans can be grown, other factors, agronomic and economic, enter into the decision to go into production.

The soybean is a major crop in the areas listed below.

1. Corn Belt States—Illinois, Iowa, Indiana, Ohio, Minnesota, Missouri, with major concentration in Illinois, southeast to northwest Iowa, south and southwest Minnesota, west and central Indiana, northwest Ohio, and northeast Missouri.

1. Background

2. Delta States—Arkansas, Mississippi, and Louisiana with major concentration in areas bordering the Mississippi River.
3. Atlantic States—North Carolina, South Carolina, Virginia.
4. Fringe States—abutting on areas 1 and 2 above. North Dakota, South Dakota, Nebraska, Kansas, Oklahoma, in each case along the eastern tier of counties; Wisconsin, Michigan, Kentucky, Tennessee, Alabama, Georgia.

More than 60% of United States soybean production now originates in the corn belt states on about 55% of the total area planted, 15% in the delta states on about 17% of the area planted, 5% in the Atlantic states, and the remainder in the fringe states.

The seven states leading in production of soybeans are Illinois, Iowa, Indiana, Missouri, Arkansas, Minnesota, and Ohio, in decreasing order. The rank order of the last four changes one place or so between years, depending on weather and anticipated prices (Table II).

In Canada the soybean is a minor crop with plantings being limited almost entirely to Ontario. Yields in Ontario have been similar to those in Ohio, on an area less than 200,000 ha (500,000 acre).

A complete analysis of the reasons underlying the rapid rise of the soybean as a major crop in United States agriculture is complex. Not all analysts agree on the weighting to be given to the factors involved. Basically, of course, financial return to the producer is paramount but compatibility with farming

TABLE II

Major Soybean Producing States in North America[a] 1960–1975

	Hectares harvested × 1000			Average yield (kg/ha)		
	1960	1970	1975	1960	1970	1975
Illinois	2014	2754	3330	1749	2085	2388
Iowa	1053	2300	2823	1715	2186	2287
Indiana	978	1328	1470	1816	2085	2210
Missouri	949	1403	1810	1446	1715	1715
Arkansas	989	1782	1904	1412	1513	1614
Minnesota	846	1227	1446	1312	1749	1749
Ohio	607	1033	1256	1648	1917	2186
Mississippi	371	1047	1264	1480	1513	1480
Canada	104	136	158	1486	2085	2327
United States[b]	9580	17110	21710	1581	1796	1910

[a] Figures converted from data in Agricultural Statistics, United States Department of Agriculture.

[b] 1 ha = 2.47 acre; 1000 kg/ha = 14.86 bushels/acre.

systems and practices of the area is also of consequence. These have undergone substantial change in the past 30 years, particularly in the corn belt, so that the context in which the soybean is judged as a profitable crop has changed also.

Because the soybean is a rich source of both oil and protein it has become the raw material for a diversity of uses, agricultural and industrial. Its economics tend to be complex and its price somewhat more volatile than that of the starchy grains. The area planted tends to fluctuate, being influenced by anticipated price in relation to other cash crops, particularly corn, carry-over stocks, export demand and weather conditions at the planting season, which, if unfavorable for corn, may result in increased soybean acreage.

Soybeans were accommodated easily in the crop sequences widely followed in the corn belt in the 1940's and 1950's. Rotations, varying in length, consisted of corn, oats, and legume hay. The nitrogen needs of the corn were supplied by the legume residual, and the phosphorus and potassium requirements, if any, were met by fertilizer application usually applied to the corn. Soybeans generally followed 1 or 2 years of corn without additional fertilizer. Land which produced acceptable corn yields seemed well suited to soybeans in the following season. Fertilizer application to soybeans seemed unnecessary except in special situations; moreover the nitrogen requirements of the beans were largely met by the symbiotic fixation process whereby nodulated legumes are able to utilize atmospheric nitrogen. It was recognized that the residual stems, leaves, and roots from a soybean crop did not enhance the soil nitrogen status as did legume-grass residues. However in much of the corn belt the practice of direct nitrogen fertilization of corn became widely adopted in the late 1950's and 1960's with near abandonment of the previous rotations including, as they did, oats and barley for which demand had decreased and prices had declined. Federal programs restricting corn acreage in the 1950's stimulated the planting of soybeans, the yields of which were rising as better adapted cultivars were developed. Soybean plantings have never been under production limitation. Because soybean planting can be delayed well after the optimum time for seeding corn with little yield penalty, soybeans are a useful replacement for corn in late wet seasons. In the delta states there has been perhaps a greater shift in cropping pattern as cotton plantings have been reduced under federal commodity programs. Although primarily the replacement cash crop for cotton, the soybean has also been substituted for feed grains in this area.

The rapid rise to prominence of the soybean in United States agriculture has not solely been due to demand for the bean and the products derived from it. Indeed both meal and oil in some years has been in surplus over current market needs. The support pricing mechanisms of the Commodity Credit Corporation, an agency of the United States government, have then

1. Background

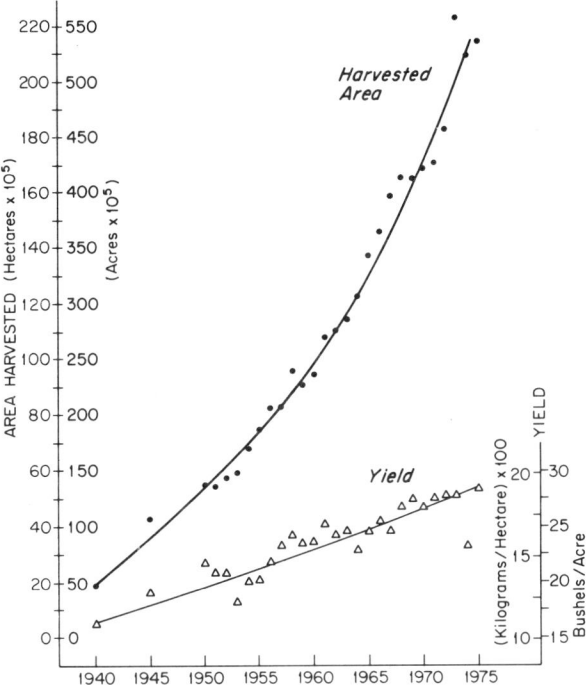

Fig. 2. United States soybean acreage harvested for beans, and average yields, 1940–1975. Prepared from Agricultural Statistics, United States Dept. of Agriculture.

come into operation. Substantial oil shipments to other countries have been made from time to time under the Public Law 480 program of foreign aid. In the 1970's the carryover of soybeans and its products from one season to the next have been small with prices on an upward trend.

The technology of soybean production in the United States has steadily advanced, as have yields. Supplementing the contribution of the plant breeder in developing new cultivars, there have been indirect effects due to the availability of better field equipment permitting timeliness of operation, and to the use of selective herbicides that have greatly reduced weed competition. Because soybean plantings have expanded into areas not as well suited as the older areas, or for which fully adapted cultivars may not yet be available, average state and national yields fall short of those attained under the best agronomic and environmental conditions. In Fig. 2 is shown the harvested area and average yields from 1940 to 1975. In some years adverse weather conditions, drought or excessive precipitation reduces yields over a substantial area. For example, prolonged drought in Minnesota and parts of

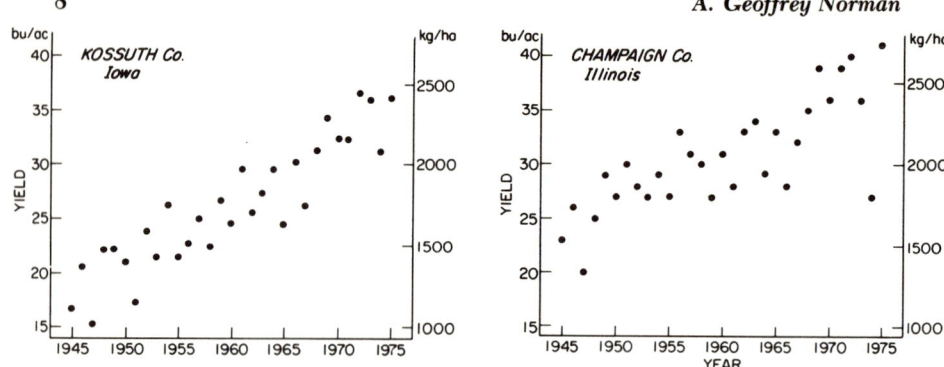

Fig. 3. Soybean acreage and yields in two high-producing United States counties, 1945–1975.

Iowa in 1976 markedly cut production in those states with an appreciable impact on the national average yield figure.

In 1975 soybean production in the United States was equivalent to a harvested yield of 1885 kg/ha (28.4 bushels/acre). The corresponding figure for 1940 was 1077 kg/ha (16.2 bushels/acre). The difference is a measure of the improvement in soybean cultivars and agronomic operations. These figures are based on weight of beans harvested and available to move into the distribution system. Only a small fraction of the soybean crop is retained on the farm by the producer either as seed or as livestock feed. Consistently 98% of the beans produced are marketed.

Overall soybean yields have moved steadily upward over the last 25 years as can be seen in Fig. 2. If one examines the yield data from a limited area with relatively uniform soils and similar farm enterprises, the integrated effect of the agronomic improvements can be estimated. In Fig. 3 are plotted soybean yield data from two high-producing counties in Illinois and Iowa where it can be assumed the best practices have been adopted. Over the 30-year-period from 1945–1975 the yield trend has been steadily upward. In Champaign County, Illinois, soybean yields in the 1970–1975 period were 30% higher than in the 1950–1955 period. In Kossuth County, Iowa, the corresponding figure was 55%. In both cases recent yields have been substantially higher than the national averages. Two counties in Illinois (Henry and Mercer) and one in Ohio (Howard) in 1975 had county-wide average yields over 2820 kg/ha (42 bushels/acre). This represents a nitrogen harvest of about 170 kg/ha. To obtain a similar nitrogen harvest in a corn crop would require substantial nitrogen fertilizer input. Furthermore, the county yields above, and experimental plot yields of 4000 kg/ha at a number of locations, suggest that there is still considerable margin for improvement in the present producing areas as better adapted cultivars become available and optimum agronomic practices are adopted.

TABLE III

World Soybean Production and Major Producing Countries[a] 1970 and 1975

	1970			1975		
	Hectarage (ha × 1000)	Production (metric tons × 1000)	Average yield (kg/ha)	Hectarage (ha × 1000)	Production (metric tons × 1000)	Average yield (kg/ha)
World	29277	41546	1437	39593	62965	1590
United States	17110	30583	1787	21710	41406	1907
People's Republic of China	8006	6900	862	8106	7200	889
North Korea	395	228	577	405	235	580
South Korea	298	232	778	274	311	1135
Indonesia	694	498	718	753	575	764
Brazil	1319	1509	1144	5747	9717	1690

[a] From Foreign Agricultural Service—Agricultural Statistics.

B. Worldwide

Although the origins of the soybean were in the Orient where this crop has long been cultivated and where it has remained a major crop to this day, production in North America now greatly exceeds that in the Orient. The main loci of soybean plantings in Asia are Manchuria, (Peoples Republic of China), North Korea, South Korea (Republic of Korea), and Indonesia, with smaller plantings in Japan and the several countries of southeast Asia. The approximate total area planted in the Orient in 1975 was something less than one-half that in North America, and the weight of beans produced was less than one-fourth. Yields per hectare are generally much lower than in the United States and seem to have changed little in the past decade, except in South Korea where an upward trend has been evident. Hectarage and yield data for these and other countries are given in Table III. It should be noted that soybeans are grown commercially in Indonesia at locations almost on the equator.

Special reference must be made to the development of the soybean crop and processing industry in Brazil which accelerated greatly in the last decade (Table IV). Between 1965 and 1975 the Brazilian soybean hectarage expanded thirteenfold with a 28% increase in yield per hectare. Brazilian soybeans now have a significant impact on world markets. Production is centered in the southern Brazilian provinces of Rio Grande do Sul, Parana, and Sao Paulo at latitudes between 20° and 30°S (Fig. 4). This corresponds in day length to Florida and Cuba in the Northern Hemisphere. The cultural practices followed are essentially those of the corn belt. Soybean penetration has been largely a replacement for wheat. Potent factors in the rapid expansion of soybean production in Brazil have been governmental subsidies, the

TABLE IV

Soybean Production in Brazil[a] 1960–1976

Year	Hectarage (ha × 1000)	Production (metric tons × 1000)	Average yield (kg/ha)
1960	172	206	1197
1965	432	523	1211
1970	1319	1509	1144
1971	1716	2077	1210
1972	2840	3666	1290
1973	3615	5400	1383
1974	4790	7400	1545
1975	5747	9717	1691
1976	6419	11227	1748

[a] From Agricultural Statistics, United States Department of Agriculture.

1. Background

Fig. 4. Soybeans from horizon to horizon in Parana, Brazil (photo by H. C. Minor).

simultaneous development of the processing and marketing infrastructure and a healthy world market for beans, oil, and meal. Furthermore, by law, bread sold in Brazil must contain not less than 5% soy flour. Additional land suitable for soybean culture is available; economics may determine the pattern of further expansion in production.

Soybeans are also planted as a minor crop in other South American countries, such as Argentina, Columbia, and Paraguay. In Europe only Romania has consistently produced a small bean crop. Soybeans are little grown in the USSR, the total harvest there in recent years being about equal to that in states such as Alabama or Nebraska.

Inasmuch as the soybean plant has a fairly wide range of adaptation, and cultivars have been developed for broad latitudinal bands it is pertinent to ask why the centers of production are as limited as they are. Considerations relating to this question are discussed in Chapter 4. Much new information about the suitability of the soybean in tropical and subtropical areas is being obtained under the INTSOY Program of the University of Illinois and the University of Puerto Rico, funded by United States Agency for International Development. A major objective is to exploit the potential of the soybean as a source of protein for direct human consumption, particularly in countries in which the diet of many is protein-deficient. The program is broad and includes the testing of improved varieties under widely different environmental and agronomic conditions, the development of new varieties that may make soybean production feasible in areas where this crop has not previously been grown and the dissemination of the technical and economic principles necessary for a soybean industry to be successful. To this end a large number of plantings have been made in many locations and countries.

These have provided a wealth of information about adaption to environment and susceptibility to diseases and pests upon which targeted breeding programs can be built. There is a high degree of probability that new areas of soybean production will appear in the subtropics.

II. USES AND ECONOMICS OF SOYBEAN PRODUCTS

The economics of the soybean industry are complex. Essentially, however, three major markets are involved: for beans, for oil, and for meal. Soybean oil is a major component of the edible oil market and is processed into a variety of products for human consumption, primarily shortening and margarine where it encounters competition from other vegetable oils. Soybean meal is in great demand as a protein source for incorporation into animal and poultry feeds. Soybeans are also processed into a wide range of products for human consumption but the amount required for these specialized uses is negligible in terms of influence on price. The price of soybeans is manifestly controlled by demand for oil and for meal, domestically and in world markets. In the United States beans ordinarily move rather promptly from the farm to local elevators and thence to large storage elevators at processing plants or export elevators in the shipment channels. In some years, notably in the late 1960's, a significant fraction of the crop was sealed in the fall under a governmental loan program intended to stabilize the price. More recently, with higher prices, soybeans have been a quick cash crop to the producer. Crushing and processing plants usually operate the year around, which requires that there be sufficient storage capacity for beans and an orderly flow of oil and meal into their respective markets.

Between 1970 and 1976 production of soybeans in the United States increased about 30% but the value of the soybean crop and the revenues from the export of beans and products more than doubled. Prior to 1970 the price of beans had remained under $3 per bushel ($82/kg) and was supported by a governmental loan program at $2.25–2.50 per bushel ($61–68/kg). In 1972 there was a substantial increase in price, briefly, to as high as $12 per bushel ($327/kg). Since then it has fluctuated considerably between $5 and $9 ($136 to $245/kg) with occasional short excursions to higher figures when in short supply toward the end of the processing year.

The United States has become the major exporter of beans to world markets and is the source of 80% of the beans moving in international commerce. In the past decade the revenue from foreign sales has been an item of consequence in the national balance of trade figures, helping to offset the increased costs of petroleum and other imports. Because of the greatly expanded production in Brazil, Brazil has now appeared as a vigorous competitor in export markets.

1. Background

In recent years about 35% of the United States soybean crop has been exported as beans, predominantly from Gulf ports, to be processed in the importing countries. More than one-half goes to countries in Western Europe, principally the Netherlands and West Germany, but the largest single importer through the years has been Japan.

The United States also exports the major soybean products, meal and oil, which are obtained by crushing and processing, as discussed in detail in the final chapter of this book. However, whole beans account for the major part of export revenues. With some variation from year to year, meal revenues constitute about one-third and oil revenues about one-tenth of the revenue from whole beans, which in 1975 amounted to $3272 million.

Soybean meal finds its primary use as a protein supplement to animal and poultry feeds. The average yield of meal from bulk soybeans is 79%. Its protein content is 50%. Soybean meal is the major source of protein concentrate for incorporation into livestock feeds and ordinarily supplies about two-thirds of the domestic requirement. Other sources supplying about 10% each are cottonseed meal, tankage and meat scraps, fish meal, and gluten meal.

Of the soybean meal produced in the United States approximately one-quarter is exported. Western Europe takes from two-thirds to three-quarters of this, West Germany, the Netherlands and Italy being the leading importers. Several countries in Eastern Europe are importers of meal. Little goes to Asia, the Orient, or South America.

Soybean oil is an edible oil used after processing in various products for human consumption. The average yield of oil from bulk soybeans is 18%. Major uses are for cooking and salad oils, shortening, and oleo margarine. There is, in addition, a small and declining amount employed in certain paints, varnishes, and resin products. As a source of oil for food products soybean oil encounters competition from other vegetable oil sources such as sunflower, palm, peanut, cottonseed, rapeseed, coconut, and olives. In recent years, however, soybean oil has filled about 70% of the United States vegetable oil requirements. The production of palm oil from large plantings in Malaysia and Africa is increasing and is influencing the export price for soybean oil.

Of the soybean oil produced in the United States only about 15% is exported. Export shipments have been from 500,000 to 600,000 metric tons annually in recent years. The distribution of the soybean oil exported is distinctly different from that of soybean meal. Western European countries meet their vegetable oil needs by processing the whole beans imported and by drawing on some of the other sources listed above. Soybean oil exports have been rather widely distributed between countries such as Iran, Yugoslavia, Mexico, Pakistan, India, Peru, Tunisia, and Morocco, with considerable fluctuation from year to year as affected by price and need. Substantial shipments have been made in connection with the Public Law 480 foreign aid program. The price of soy oil tends to be rather more volatile than that of meal or beans and to be more influenced by carryover stocks.

The uses of the two major soybean products, meal and oil, are unrelated, and their price in the market tends to move independently, subject to the obvious supply limits set by the protein and oil content of the soybean. As the export markets for beans, meal and oil have expanded, purely domestic supply and demand considerations have had less influence on price. From one crop season to the next the carryover stocks of beans and of products have in recent years been relatively small. There have, on occasions, been shortages of soybeans in the late summer before the new harvest moves into supply channels.

In the preceding paragraphs attention has been directed towards the oil and meal uses because these are large and constitute the reasons for the development of the soybean as a major crop in the United States. There are, however, a diversity of other uses, mostly in the food industry, which do not in the aggregate account for more than a small fraction of the total crop and have a negligible influence on price. The food industry is increasingly using products derived from soybeans as additives to manufactured food products. Sometimes the supplements are intended to increase the content of protein or oil but more often to influence physical structure, stability, or texture. Flavor is not an objective; on the contrary the products incorporated are processed to be tasteless. Soybean flour, either full fat flour or defatted flour, is added to some bakery goods. Soyprotein concentrate in various forms is incorporated in some meat products as an extender, but also in a textured form to simulate meat. Lecithin, a phosphatide removed from crude soybean oil by aqueous extraction, finds uses as a stabilizer and humectant.

III. THE SOYBEAN IN PHYSIOLOGICAL RESEARCH

Although this book deals primarily with the soybean as a crop plant, it is worth noting that the soybean has been the experimental plant of choice in a diversity of physiological studies. It is rapid in growth. Most cultivars do well in greenhouses or controlled environment chambers if the light level is adequate. A high degree of uniformity between individual plants can be obtained by selecting quality seed of equal weight from a pure cultivar lot. Plants may be grown satisfactorily in soil, in nutrient solution, or in gravel culture. Attention has to be given to day length and light intensity if there is a requirement that studies at different times of year be directly comparable. Good predictions of probable yield under field conditions cannot be made from plants grown in the greenhouse or controlled environment rooms.

1. Background

In studies on the initiation of flowering, the soybean has frequently been used as a representative of the class of short-day plants. The onset of flowering is mediated by exposure to a night or dark period in excess of a critical length (see Chapter 2). The soybean cultivar BILOXI was the plant of choice in classical studies on the photoperiodic response.

The leaf arrangement on the stem of the soybean makes it convenient for experiments on assimilation and translocation, much aided by the use of isotope tracers, radioactive, such as ^{14}C and ^{32}P, or stable, such as ^{15}N. The leaf is the site of photosynthesis. Export of photosynthate from the leaf to the roots or to the developing shoot can therefore be followed by isotopic labeling. Similar procedures have been adopted in the study of the maturation process of pod-filling, when leaf components including mineral nutrients are depleted by transport to the seed.

The soybean has been of especial significance in researches on the biochemistry of the process of symbiotic nitrogen fixation in the root nodules which are developed on plants in the presence of *Rhizobium japonicum* (see Chapter 2). Soybean nodules are large and not as convoluted in form as those developed on most pasture legumes. Surface sterilization of the nodules is easily accomplished if nodular tissue is required. Nitrogen fixation by excised soybean nodules has been demonstrated for a limited period after removal, again with the aid of isotopic nitrogen. Soybean nodules have proved to be a good source of the respiratory pigment, leghemoglobin, which is essential to the fixation process and unique to the nodule, not being found either in unnodulated root tissue or in the symbiotic bacteria.

In greenhouse and field experiments the soybean has been used in attempts to understand the relationship between available soil nitrogen and the amount of nitrogen fixed by the symbiotic mechanism. Maximum yields are not obtained if no fixed nitrogen is present in the root zone, but fertilizer nitrogen additions depress the contribution made by the nodular mechanism (see Chapter 3).

The paragraphs above inadequately outline a few of the characteristics of the soybean that render this plant of interest to the experimental biologist as well as to the agronomist whose attention is centered on its improvement and culture as a productive crop.

2

Growth and Development

D. R. HICKS

I.	Plant Development	17
	A. Introduction	17
	B. Seed Morphology and Germination	18
	C. Rooting Pattern, Depth, and Duration	19
	D. Nodule Initiation and Structure	20
	E. Shoot Morphology and Developmental Pattern	22
	F. Canopy Development	28
II.	Environmental Effects on Plant Development and Performance	30
	A. Moisture	31
	B. Temperature	33
	C. Day Length and Light Intensity	35
	D. Soil Aeration	37
	E. Soil Condition and Cropping Practices	37
	F. Soil Fertility	39
	G. Exogenous Growth Regulators	41
	References	41

I. PLANT DEVELOPMENT

A. Introduction

The soybean, *Glycine max* (L.) Merrill, is an annual plant commercially grown primarily for oil and protein production. Even though morphological diversity exists, the soybean is generally a plant which grows 90–120 cm in height with the first leaves simple and opposite and all other leaves alternate and trifoliolate. All above ground vegetative parts are covered with many, small hairs. Branches may develop from buds in the lower leaf axils. Flowers develop in all leaf axils which give rise to 0–5 pods per node with 1–5 seeds per pod. Plants are either determinate or indeterminate in flowering and developmental habit.

The developmental morphology of the soybean has been extensively described by Carlson (1973). The objective of this chapter is to summarize morphological development and to emphasize the effect of the environment on plant development and performance.

B. Seed Morphology and Germination

The seed of the soybean varies in shape, but is generally oval and consists of an embryo enclosed by the seed coat. Very little endosperm tissue exists. The seed scar (hilum) is easily visible on the external surface of the seed coat (Fig. 1). The hilum is oval in shape and occurs when the seed breaks away from the ovary. The micropyle is a small hole in the seed coat formed during seed development and is located at one end of the hilum. The hypocotyl–radicle axis located above the micropyle is sometimes visible through the seed coat. The seed coat consists of eight to ten layers of cells—the outer layer of palisade cells is the epidermis, the hypodermis is the next layer of loosely packed cells, and finally the inner parenchyma tissue which is six to eight layers of thin-walled, flattened cells (Dzikowski, 1936). Gas exchange between the embryo and its external environment occurs largely through the micropyle because of the cutinized outer layers of cells. However, water is absorbed through the entire seed coat surface.

The embryo is comprised of two cotyledons, a plumule with the two simple leaves, and the hypocotyl–radicle axis.

Germination is a complex metabolic and physiologic process that begins with the seed and results in a plant capable of completing a normal life cycle. Under favorable environmental conditions, food reserves in the seed are utilized to develop the root and shoot, the seedling emerges from the soil

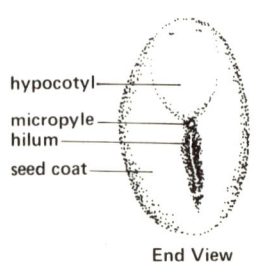

Fig. 1. Mature soybean seed.

and becomes self-sufficient, and the plant perpetuates the species by developing viable seeds for the next generation.

The radicle emerges from a break in the soybean seed coat near the micropyle and begins to grow downward within 1–2 days after seeding (Williams, 1950). Branch roots begin to develop when the radicle is 2–3 cm long and anchor the seedling in the soil. The hypocotyl arch is the seedling structure which breaks through the soil surface and straightens soon after emergence. Chlorophyll development in the cotyledons is rapid after emergence and the seedling begins to supplement the cotyledonary energy reserves with the onset of photosynthesis.

C. Rooting Pattern, Depth, and Duration

Maximum soybean seed yield depends to a large extent upon a well-nodulated, extensive root system, the development of which is enhanced by ample water and nutrients from the soil and energy from respired assimilates. Soybean cultivars and strains differ in root surface area and root dry weight.

The soybean was first characterized as having a taproot with many branches which penetrated to a depth of 150 cm with the major portion occurring in the upper 60 cm (Borst and Thatcher, 1931). Recent reports (Mitchell and Russell, 1971; Raper and Barber, 1970) indicate that field-grown soybeans lack a distinct taproot, with a major portion of the root system consisting of lateral roots arising from the upper 10–15 cm section of the primary root. These lateral roots extend outward from the plant nearly horizontally for 40–50 cm and then grow downward to depths as great as 180 cm.

Root depth increases faster than shoot height during vegetative development. Root depth is almost two times shoot height until reproductive growth begins (Mayaki *et al.*, 1976). Even though root depth may exceed plant height throughout most of the growing season, the dry weight of the above ground plant parts exceeds the root dry weight throughout the season; the shoot:root ratio constantly increases under normal growing conditions (Roberts and Struckmeyer, 1946; Earley and Cartter, 1945). Root growth continues until sometime during the seed filling period when it declines and ultimately ceases prior to seeds reaching physiological maturity.

The root tip consists of three regions: the promeristem, primary meristem, and primary permanent tissues. From the promeristem and primary meristem tissue located at the root tip, more mature and permanent tissues develop. The length of the promeristem and primary meristem is about 300 μm. The permanent tissues ultimately consist of xylem, phloem, pericycle, endodermis, cortex, and epidermis. In a cross section of the root 3 cm from the root apex, the tetrarch xylem pattern is visible.

Secondary roots emerge from the primary root; their origin is the pericycle tissue opposite the ridges of the xylem. They first develop 4-5 cm from the root apex (fourth to fifth day of seedling age) and continue to develop acropetally as the primary root continues to grow (Anderson, 1961; Sun, 1955). The secondary root is usually smaller in diameter than the primary root. Tertiary and higher orders of roots form from secondary roots which continues throughout the life of the root system. The structure of roots of higher order is similar to the primary root except that they may be triarch and diarch, that is, they lack one or two xylem ridges and phloem strands.

The root epidermis consists of thin-walled cells without intercellular space. Epidermal cells give rise to the root hairs which appear about 1 cm from the tip of the primary root 4 days after germination (Anderson, 1961). Root hairs also form on secondary and higher order roots as the root system develops. Dittmer (1940) found root hairs on all surfaces of the root system except the taproot of mature, field-grown ILLINI soybeans. Root hairs which were developed during the early growing season appeared intact at the time of harvest.

From root hair number and length of 14-week-old greenhouse-grown OTTAWA-MANDARIN soybeans, Carlson (1969) calculated the surface of the root system to be about 1.2 m^2. The major portion of the roots were tertiary or higher order roots which also bore the major portion of the root hairs. Although data apparently are not available, it is reasonable to assume that the root surfaces of nonstressed, field-grown soybean plants are greater than 1.2 m^2.

There is evidence that the root system may be a yield-limiting factor in soybeans. Sanders and Brown (1976) varied the shoot:root ratio of soybeans to determine its effect upon the growth of soybeans. Seven-day-old LEE 68 seedlings were grafted such that the shoot:root ratios were 1:1, 1:2, 1:3, and 3:1. Grafted plants were field-grown to maturity. Seed yield and number of seeds per plant were increased by multiple shoots, but multiple roots caused an even greater seed yield and number of seeds per plant. Leaf area and leaf number per plant, determined during the early reproductive growth stage, were increased by multiple shoots and, increased to a greater extent by multiple roots. At this developmental stage, however, the dry weight shoot:root ratio was nearly the same (approximately 10:1) for all grafting treatments.

D. Nodule Initiation and Structure

Nodules may form on soybean roots after root hairs are present and are initiated by *Rhizobium japonicum*. Roots secrete substances which promote rapid growth of soil microbes, including nodule bacteria (Dart and Mercer,

2. Growth and Development

1964; Nutman, 1959). Early investigators postulated that the highly flagellated cells of *Rhizobium* entered root cells by inducing an increase in pectic enzymes which softened the cell wall by partial dissolution (Ljunggren and Fahraeus, 1961; Raggio and Raggio, 1962). Others have reported that pectic enzymes did not significantly affect the initial infection process (Lillich and Elkan, 1968; Macmillan and Cooke, 1969).

Regardless of the mechanism of entry, bacteria may enter the soybean root via the root hair cell or other epidermal cells (Carlson, 1973). The first sign of infection is an elongation and curling of the tip of the root hair. After entry of bacteria, cytoplasm of the host cell forms an infection thread (Goodchild and Bergersen, 1966). This infection thread grows 60–70 μm (to the base of the epidermal cell) in about 2 days (Bieberdorf, 1938). It continues to grow inward through the cortex of the primary root and may grow completely through the cortex of secondary and higher order roots, but never penetrates the endodermis or pericycle of these smaller roots.

Most infection threads do not induce nodule formation. They may fail to penetrate beyond the root hair or are unsuccessful in stimulating nodule formation even if they penetrate some cortical cells (Raggio and Raggio, 1962).

The infection thread may branch as it extends through the cortex, thereby infecting many host cells. The physiological stimulation of the cortex cells to initiate nodules by the infection thread is not understood. Cortex cells of some genera of legumes (*Lathyrus, Lespedeza, Medicago, Pisum, Trifolium,* and *Vicia*) have nuclei with twice the normal chromosome number; these disomatic cells may be the site of nodule initiation (Wipf and Cooper, 1940; Vincent, 1962). Wipf (1939) stated that for these genera, a tetraploid cell must be intercepted by the infection thread which becomes the nodule initial. The disomatic cell site of nodule initiation hypothesis has not been supported in soybeans. However, Carlson (1973) observed tetraploid cells in 7-day-old nodules of OTTAWA–MANDARIN soybeans, but found no disomatic cortical cells prior to nodule formation.

The infection thread serves as an entry mechanism for one or more bacteria which undergo rapid division for 2 weeks (Bergersen and Briggs, 1958). The host cells concomitantly rapidly divide causing breakage and disappearance of the infection threads (Bieberdorf, 1938). After 2 weeks of cell division, further growth of host cells is due to cell enlargement in an acropetal direction. Diameter growth ceases by the end of the fourth week after nodule initiation, resulting in mature nodules 3–6 mm in diameter, irregular in shape, and containing a vascular system continuous with the host root.

The bacteria also divide rapidly at this time so that the cells in the central region of the nodule become filled with bacteria, known at this stage as bacteroids. Bacteroids, in groups of four to six, become enclosed in membrane envelopes (Goodchild and Bergersen, 1966) which apparently also contain leghemoglobin (Bergersen and Goodchild, 1973).

The characteristic pink coloration of healthy nodules is due to leghemoglobin which increases during the first.2 weeks after nodule initiation. Nitrogen fixation begins with the appearance of leghemoglobin and cessation of bacterial division. Fixation continues until the sixth or seventh week of nodule age when nodule senescence begins (Bergersen, 1958).

Nodule initiation and development is a continuing process as root development continues. Therefore, nodules of all ages may be present on a mature soybean plant.

There is a marked specificity between the bacteria and the host at the time of initial infection. As a result of this specificity, a given species of legume can only be infected by a given species of *Rhizobium*. For soybeans, the infective species is *Rhizobium japonicum*. The specificity between the host and the *Rhizobium* is apparently first expressed at the time of infection thread formation (Li and Hubbell, 1969).

Hamblin and Kent (1973) first provided evidence that plant lectins may be involved in the *Rhizobium*–legume specificity by showing that lectin from *Phaseolus vulgaris* seed could bind with *Rhizobium phaseoli*, the symbiont for beans. Plant lectins are proteins which are capable of binding to certain sugars or sugar-containing proteins with a high degree of specificity. They may constitute from 2–10% of legume seed proteins and smaller amounts of the protein in legume leaves, stems, and roots (Liener, 1976).

Bohlool and Schmidt (1974) tested the ability of soybean lectin to bind with twenty-five strains of *Rhizobium japonicum* and twenty-three strains of other *Rhizobium* species which are not infective on soybean roots. They found that soybean lectin would bind with twenty-two of the twenty-five *Rhizobium japonicum* strains, but with none of the twenty-three strains of other *Rhizobium* species. More recent evidence indicates that the host–*Rhizobium* specificity is not determined by simple attachment of the *Rhizobium* to the legume root (Chen and Phillips, 1976). It thus remains to be seen precisely what the biochemical basis for specific recognition between the *Rhizobium* and the host might be, and what role lectins may play in this recognition.

E. Shoot Morphology and Developmental Pattern

The development of above ground parts of the soybean plant begins with emergence of the hypocotyl from the soil and terminates with the formation of mature seed. Above ground soybean structures are diagrammed in Figs. 2, 3, and 4. Under favorable conditions for growth, the plant emerges 4–7 days after planting. Plant dry weight increases at a slow rate initially and

2. Growth and Development

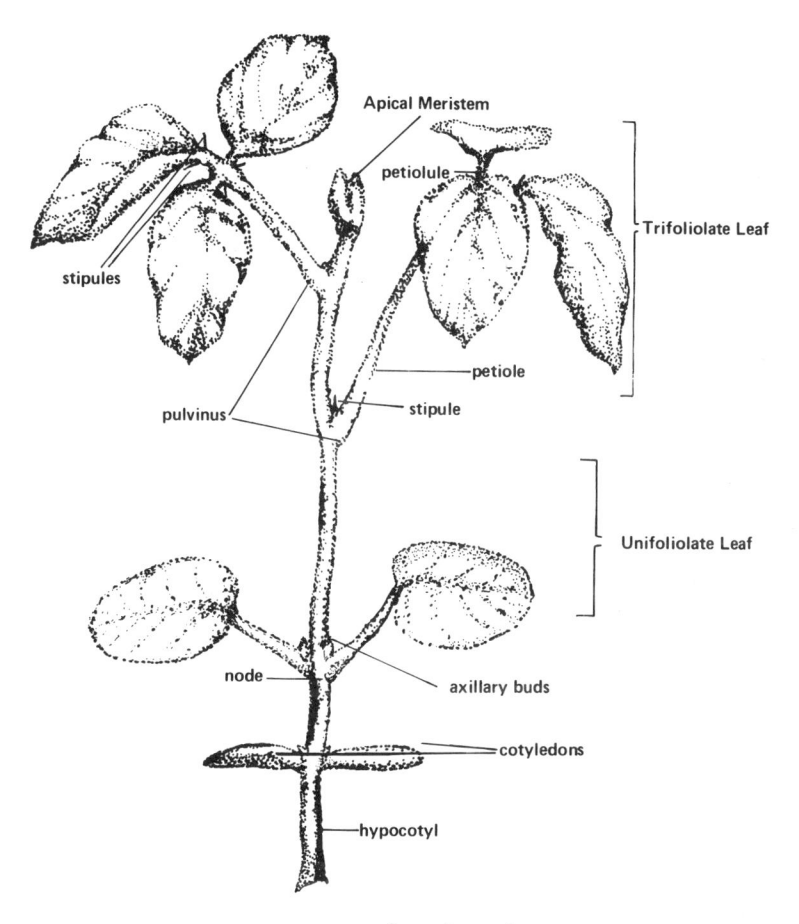

Fig. 2. Structure of a soybean plant.

then more rapidly during a period where the increase is linear with time. The late growth phase is again slower as the plant approaches physiological maturity. Vegetative growth is almost complete with the onset of seed enlargement which marks the beginning of a period where dry weights of leaves, stems, and roots decrease slightly.

The mature soybean plant has 19–24 nodes which are completely differentiated in 4–5 weeks after planting (Johnson *et al.*, 1960). The lowermost node is the point of attachment of the cotyledons, the next node gives rise to the opposite unifoliolate leaves, and all subsequent nodes produce single trifoliolate leaves alternately up the stem.

Floral primordia are initiated within 3 weeks and flowering begins within 6–8 weeks after emergence when grown in an area of adaptation. Pods are

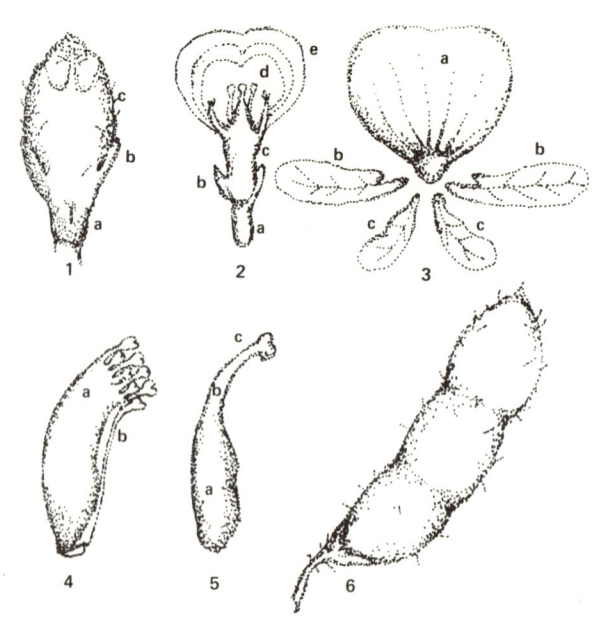

Fig. 3. The soybean floral bud and its parts. (1) Young bud. a. pedicel, b. bract, c. calyx. (2) Flower. a. pedicel, b. bract, c. calyx, d. stamens, e. petals. (3) Part of corolla. a. standard, b. wings, c. keelpetals. (4) Androecium. a. nine united stamens, b. one free stamen. (5) Pistil with ovary. a. ovary, b. style, c. stigma. (6) Mature pod.

visible in 10 days to 2 weeks after the onset of flowering. Flowering continues for 3–4 weeks; many stages of pod and seed development occur on the plant until near physiological maturity. Seeds in all pods mature within a period of about 1 week even though pollination times may have differed greatly.

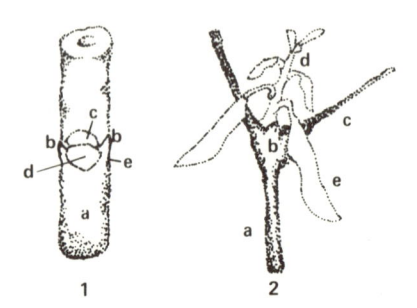

Fig. 4. A soybean axillary raceme. (1) Portion of main stem. a. stem, b. stipule, c. raceme scar, d. leaf scar, e. node. (2) Stem section with an axillary raceme. a. stem, b. stipule, c. petiole, d. axillary raceme, e. pod.

1. Stem

The embryo axis in the seed consists of root and shoot tissue; the shoot comprises the hypocotyl and the epicotyl. The epicotyl consists of two simple leaves, the primordium of the first trifoliolate leaf, and the stem apex. Therefore three nodes are present in the mature embryo, the cotyledonary, unifoliolate, and first trifoliolate nodes. The four protoxylem poles continue from the root up to the top of the hypocotyl where they apparently merge into two, one for each cotyledon. The primary xylem of the root–hypocotyl–cotyledon axis is a continuous unit (Weaver, 1960).

The shoot apex consists of a two-layered tunica and a massive corpus. The corpus has three distinct zones including a central initiation zone of large cells, a peripheral zone of smaller cells, and a rib meristem immediately below the initiation zone. The initiation and development of the second and subsequent trifoliolate leaves occur 30–50 μm below the summit and to the side of the stem apex. Leaf development occurs alternately up the stem (Sun, 1957; Decker and Postlethwait, 1960).

The plastochrome, the time interval between initiation of leaves, in soybeans is about 2 days. Miksche (1961) found it took about 3½ days after germination for the second trifoliolate leaf to be initiated in HAWKEYE soybeans, and subsequent plastochromes required about 2 days. Johnson *et al.* (1960) counted nineteen nodes on the main axis 35 days after planting which supports the plastochrome interval of about 2 days in soybeans.

The mature stem consists of an endarch, collateral eustele, pericycle, cortex, and epidermis. The procambial strands continue up the stem with continued development; divisions of the procambium strands give rise to the phloem and xylem elements. The mature stem is largely pith tissue (large parenchyma cells) and is sometimes quite woody in the lower section.

2. Leaf

Mature leaves consist of epidermis, mesophyll, and a vascular system. Both the upper and lower-epidermal layers are covered with a thin cutin layer (Williams, 1950). Stomata are present on both surfaces; there are about three times as many on the lower epidermis as on the upper epidermis (Carlson, 1973). Stomata are bordered by guard cells about 6 μm in width by 24 μm in length. The mesophyll consists of two layers of palisade parenchyma cells and two to three layers of spongy parenchyma. All mesophyll cells contain chloroplasts, but the two layers of palisade cells contain the majority. The vascular system of the stem continues into the petiole, the leaf midrib, and lower order veins of the leaf.

There are four types of soybean leaves: the cotyledons, primary leaves, trifoliolate leaves, and the prophylls. The cotyledon (seed leaf) is semicircular in shape enclosed by an epidermis with stomata on both the upper and lower surfaces. The interior mesophyll consists of palisade tissue with more spongy parenchyma cells which lack distinct layer formation on the abaxial side. The cotyledons function as an energy reserve until the plant becomes autotrophic. They turn yellow and drop from the plant during early vegetative development.

The primary or unifoliolate leaves are opposite at the node immediately above the cotyledonary node. They have petioles 1–2 cm long and have a pair of stipules at the point of petiole attachment to the stem. All other leaves produced on the plant (both main stem and branches) are trifoliolate and alternately arranged. Leaflets range from oblong to ovate to lanceolate in shape and have entire margins. Leaflets of the mature trifoliolate leaf generally vary from 4 to 20 cm in length and from 3 to 10 cm in width. A narrow leaflet trait occurs in some varieties due to the homozygous condition of the *na* gene (Woodworth, 1932). Williams (1950) reported deviation from this leaflet pattern in that occasionally the compound leaves consist of four to seven leaflets and fusion of the lateral with the terminal leaflets may occur.

A large pulvinus exists at the point of attachment of the petiole to the stem for each primary and trifoliolate leaf. A smaller pulvinus also occurs at the point of attachment of each leaflet to the petiole. Changes in osmotic pressure in various parts of the pulvini cause the leaflet positions to change during night and day.

The prophylls are very small, simple leaves which occur in pairs at the base of each lateral branch and the lower part of the pedicel of each flower. They lack petioles and pulvini.

Pubesence due to epidermal hairs or trichomes is present on the stem, leaf, sepals, and pods of most soybeans. Glabrous lines exist but are not used as commercial varieties. Trichomes vary in size, color, density, and form (Bernard and Singh, 1969). They consist of a single cell 20–30 μm in diameter and from 500 to 1,500 μm in length which is liquid-filled on young leaves and ultimately becomes air-filled or flattened.

3. *Flowering*

Flower initiation is controlled by photoperiod, temperature, and genotype. Soybean stem growth and flowering habit are of two types: indeterminate and determinate. The indeterminate type is characterized by the apical meristem continuing vegetative activity during most of the growing season; the infloresences are axillary racemes, and pods are produced rather uniformly (comparable number at nodes) up the stem. The determinate stem type is characterized by vegetative development which ceases when the apical meristem becomes an infloresence, both axillary and terminal racemes exist, and pods are borne rather uniformly along the stem except for the cluster of pods at the terminal raceme. The flowering period and the time of overlap of vegetative and reproductive growth are greater for the indeterminate than the determinate type.

Whether the uppermost raceme is axillary or terminal has not been anatomically demonstrated. Bernard (1972) used the terms "abruptly terminated" (determinate) and "tapered" (indeterminate) and suggested the difference between the two types is a result of the timing of the termination of stem growth rather than the way growth terminates. Two genes, dt_1 and Dt_2, control stem type. The homozygous dt_1 conditions the determinate type and the heterozygote an intermediate or semi-determinate type. The Dt_2 gene results in an intermediate or semi-determinate type.

Dry matter accumulation patterns in determinate and indeterminate lines were found to differ (Egli and Leggett, 1973). With the onset of flowering, the determinate line had grown to 84% of its maximum height compared with only 67% for the indeterminate variety. Of the total above ground dry weight (not including seed), the indeterminate had produced 58% when flowering began and 87% when pod development began. The determinate line had produced 78 and 92% of its above ground nonseed dry weight with the onset of flowering and pod developmment, respectively.

The position of the node producing the first flower depends upon the stage of development when floral initiation occurs. The nodes of the cotyledons, the primary leaves, and the first trifoliolate are differentiated in the mature embryo. Therefore, the node producing the first flower must be node four (second trifoliolate leaf node) or above. Unexpanded buds in the embryo can produce flowers after initiation, but are vegetative buds initially. Floral primordia are visible microscopically within 3 days after induction. Borthwick and Parker (1938) observed that flowers were first evident at the node which was second from the stem tip when floral initiation was imposed. Flowers develop progressively up and down the main stem from the first flowering node and outward toward tips of lateral branches.

Flower clusters may contain 2–35 flowers. Extensive flower abortion (20–80%) can occur at any stage of development from the time of bud initiation to seed development (Hardman, 1970; VanSchaik and Probst, 1958). The most frequent time for flower and/or pod abortion is 1–7 days after flowering (Kato et al., 1955). In addition, individual ovules within an ovary may abort. The basal and terminal ovules abort most frequently.

Because of the structure of the floral parts, soybeans are almost completely self-fertilized. Pollination may occur within the bud or before the flower completely opens. Crossing has been estimated to be less than 0.5% (Caviness, 1966). Lack of fertilization probably has little effect on the high percentage of flower abortion in soybeans and is not thought to limit seed yield.

However, Erickson (1975) recently reported an increase in seed yield as a result of placing bee cages in soybean fields.

4. Seed Development

Seed development after fertilization is rapid. Within 7 days after fertilization the cotyledons are initiated, tissue systems of the hypocotyl are defined within 12 days, primordia of the primary leaves are formed after 14 days, cotyledons reach maximum size in 26 days and primary leaves in 30 days, and the first trifoliolate leaf primordium is differentiated in 30 days. Seeds are physiologically mature in about 65–75 days and contain about 55% moisture at physiological maturity (Delouche, 1974). Mature pods may contain one to five seeds with two to three seeds per pod most common. Pod length varies from 2 to 7 cm and may be light yellow, yellow-gray, brown, or black in color.

Seed dry matter accumulation rates have been reported to be 80–130 kg/ha/day (Egli and Leggett, 1973; Hanway and Weber, 1971a).

F. Canopy Development

1. Dry Matter Production and Distribution

Dry matter accumulation in various above ground plant parts is shown in Figs. 5 and 6 (Hanway and Weber, 1971a). Between stages 5 and 9, the

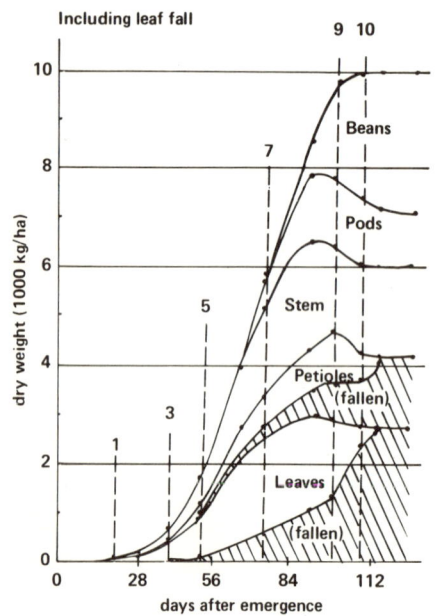

Fig. 5. Cumulative dry weights of soybean plant parts at various growth stages including fallen leaves and petioles. (Reproduced from *Agronomy Journal*, Vol. 63, pp. 263–266, 1971 by permission of the American Society of Agronomy).

2. Growth and Development

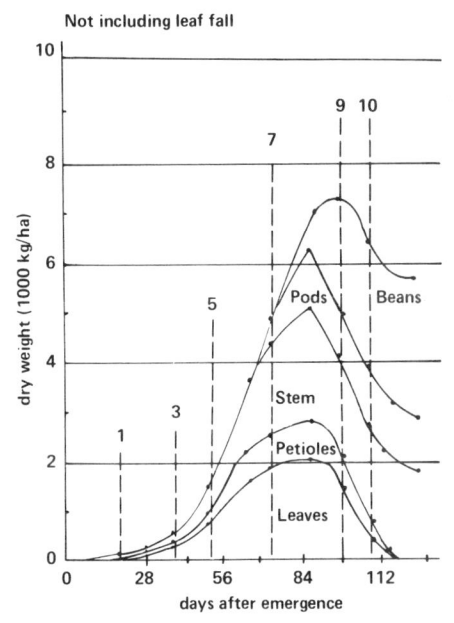

Fig. 6. Cumulative dry weights of soybean plant parts at various growth stages, not including fallen leaves and petioles. (Reproduced from *Agronomy Journal*, Vol. 63, pp. 263-266, 1971 by permission of the American Society of Agronomy).

average daily increase in dry weight was 186 kg/ha/day. Similar dry weight increases have been measured for determinate soybean varieties (Egli and Leggett, 1973). Increase in seed yield becomes rapid at about stage 8 while other plant parts lose weight after stages 8 and 9. At maturity, above ground dry matter consists of approximately 28% leaves (fallen), 15% petioles (fallen), 17% stems, 11% pods, and 29% seed.

2. Growth Analysis

Growth of plant communities has been studied by a technique called "Growth Analysis" whereby certain calculations are made relative to the plant material present and the assimilatory area during the growing season.

Crop growth rate (CGR) is defined as the increase of plant material per unit of time (Radford, 1967) and is a function of net assimilation rate (NAR) and leaf area index (LAI). NAR is the increase of plant material per unit of assimilatory material per unit of time and LAI is the leaf area of the plant material per unit area of ground.

CGR increases for the first 50-60 days after soybean planting and declines sharply thereafter (Buttery, 1969, 1970). NAR declines throughout the season as LAI increases because leaf development gives rise to self shading of lower leaves. Lower leaves are older and less active photosynthetically which also contributes to the decline in NAR.

Dry matter production is related to LAI. For some crops an "optimum" LAI exists such that the rate of dry matter production is at a maximum at a particular LAI and is less at LAI values below or above this value (Donald, 1963). For those crops with an optimum LAI, lower leaves are apparently parasitic and require photosynthates for maintenance by translocation from active upper leaves. Soybeans do not exhibit an optimum LAI (Weber et al., 1966; Buttery, 1969). Dry matter production increases as LAI increases up to about 5.0 but does not decrease at greater LAI values.

High plant populations and narrow row spacings favor attainment of maximum LAI's. However, maximum LAI's may not always give rise to maximum seed yield. Weber et al. (1966) reported maximum seed yield occurred at LAI less than the LAI for maximum dry matter production.

The percentage solar radiation intercepted by soybean canopies increases as LAI increases and the percentage interception is linearly related to increased dry matter production (Shibles and Weber, 1966).

Relative growth rate (RGR) is defined as the increase of plant material per unit of material present per unit of time. RGR is a function of NAR and leaf area ratio (LAR) where LAR is the ratio of the assimilatory material per unit of plant material. RGR is a measure of the plants efficiency in increasing dry weight with a given amount of assimilatory material at a given point in time.

RGR of each individual plant fraction increases, at a decreasing rate, as the season progresses (Koller et al., 1970). At any given time, the most recently initiated plant fraction has the greatest RGR. For all plant fractions, RGR increases to a peak and then declines. Growth rate of the leaf component peaks first, followed by that of the stem and petiole, pod wall, and seed. RGR for each individual plant fraction for branch, lower main stem, middle main stem, and upper main stem shows developmental patterns within the soybean community to be similar to the whole plant growth pattern (Koller, 1971). The lower main stem produces the most leaf area which abscises so that the middle main stem segment contains the most leaf area at the time of rapid seed development. The time interval between vegetative and seed development decreases toward the top of the plant which results in more overlap of vegetative and seed development growth curves. Seed RGR does not vary due to position within the soybean plant.

Specific leaf weight (SLW), the ratio of leaf weight to leaf area, and NAR have increased as selection for yield in soybeans has been practiced (Buttery and Buzzell, 1972) suggesting that these are traits which could be used as selection indices and lead to increased soybean seed yield.

II. ENVIRONMENTAL EFFECTS ON PLANT DEVELOPMENT AND PERFORMANCE

The soybean plant responds to its environment by changes in its development and function. The following paragraphs consider the effects of moisture, air and soil temperature, day length and light intensity, soil aeration, soil physical condition and cropping practices, soil fertility, and exogenous growth regulators on soybean development.

A. Moisture

1. Germination and Root Growth

Maximum percentage emergence occurs under optimum conditions of soil moisture, temperature, gases (oxygen), and soil physical condition.

The first process in the germination sequence of events is imbibition of water. Water availability is, therefore, the major environmental factor affecting germination. The seed moisture content (dry weight basis) required for soybean germination is about 50% compared with 30% for corn, 26% for rice, and 31% for sugar beets (Hunter and Erickson, 1952). To support soybean seed germination, the soil moisture level must be higher than that for corn. Soil moisture tension cannot be less than −6.6 bars for soybean germination to occur within 5–8 days at 25° C while tension can be as great as −12.5 for corn germination.

Soil water is depleted to the same depth as root penetration during the early part of the growing season while later in the season water is depleted to a depth 15 cm greater than root penetration (Stone et al., 1976). A small portion of the root system may be responsible for much of the water uptake (Reicosky et al., 1972).

Excessive soil moisture levels are not conducive to germination and early root growth of soybeans. Soybean roots develop faster at soil moisture tension of −0.5 bar than at −0.9 bar (Grable and Danielson, 1965). Because oxygen is required for root respiration, excess soil moisture may limit available oxygen, which in turn slows root growth and also provides an environment that fosters the occurrence of root rot organisms such as *Phytophthora megasperma* var. *sojae* (Hildebrand, 1969).

2. Shoot

Water needs of the soybean increase as the plant grows. Water lost by transpiration during the day frequently exceeds that absorbed by the roots, which creates a diurnal, internal water stress (Slatyer, 1967). This stress can occur when water is available in the soil but it is, of course, enhanced by soil water deficit and by hot, dry winds which increase the evaporative demand on the plant. Any period of stress, regardless of length, would be expected to cause changes in the plant which reduce metabolic activity, and hence, lower seed yield.

Plant height, number of nodes, stem diameter, number of flowers, percentage of pod set, number of seeds, and seed weight are positively related to soil moisture (Ueda, 1952).

The water supply to the leaf for maintaining turgor pressure in developing cells is an important factor in determining the rate of leaf enlargement (Lockhart, 1965). Rapid leaf enlargement is important because leaves are the major light-intercepting and photosynthesizing organ of the soybean. Shibles and Weber (1965) found leaf area and growth rate of soybeans to be nearly proportional. Therefore, early moisture stress reduces growth rate by inhibiting leaf enlargement. Boyer (1970) found that the rate of soybean leaf enlargement was rapid with leaf water potentials greater than -4 bars and rapidly declined with potentials between -4 and -12 bars. Growth rate was negligible at -12 bars. Field-grown soybean plants which had been water-stressed for approximately 60 days had a smaller leaf area than nonstressed plants (Ciha and Brun, 1975).

Moisture stress causes maximum reduction in seed yield if it occurs during the pod-filling stage (Shaw and Laing, 1966; Doss et al., 1974). Seed yields were reduced by all stress treatments. Water stress during flowering and early pod development caused more flower and pod abortion whereas seed size was reduced by water stress during the later stages of pod filling.

Soybean genotypes differ in the size of the root system. HAROSOY 63 has a root surface area almost twofold greater than AODA (Raper and Barber, 1970). Sullivan and Brun (1975) grew CHIPPEWA 64 scions grafted onto these rootstocks to study the effect of root volume on scion performance when plants were stressed at various growth stages. Water stress symptoms of scions on AODA rootstocks were more severe than of scions on HAROSOY 63 and CHIPPEWA 64 rootstocks. During pod filling, scions on AODA rootstock had higher stomatal resistances and lower leaf water potentials and photosynthetic rates than scions on the other rootstocks. Root volume was not measured, but presumably these differences were due to greater volume rather than more efficient water uptake and more rapid internal water movement. Seed yield was reduced for all stressed versus nonstressed plants; the greatest reduction occurred with stress during pod filling. Fewer pods per plant occurred when stress was applied during flowering. Seed weight was lower when stressed during preflowering or pod-filling stages.

Even though yield may be reduced by stress, the soybean plant can withstand a short period of drought without greatly affecting yield because of its ability to alter the proportion of flowers which abort at each node. Hicks and Pendleton (1969) removed all flowers from one-third and two-third sections of field-grown plants and found pod number per node increased at nodes in the untreated section of the plant. All flowers might abort at some nodes because of moisture stress while fewer abort at subsequent nodes, providing the stressed conditions have been eliminated or lessened. As a result, the plant is affected less by drought than is a crop such as corn which flowers over a shorter time period.

Soybean genotypes differ in their response to low soil moisture. Mederski *et al.* (1973) measured the seed yield of eight cultivars grown under conditions of low and high soil moisture. Low soil moisture plots were nonirrigated and received less than half the normal July–August precipitation while high soil moisture plots were irrigated to maintain available soil moisture at or above 80%. All cultivars performed best when grown under high moisture conditions. However, some later maturing cultivars performed nearly as well under low moisture conditions as they did under high moisture conditions. The authors concluded the susceptibility of soybeans to moisture stress was probably related to maturity.

B. Temperature

1. Germination

Maximum germination of soybeans in the shortest time occurred at a constant temperature of 30°C with no improvement in germination by alternating the temperature (Delouche, 1953). Inouye (1953) reported the optimum temperature for germination to be 34°–36°C, the minimum as 2°–4°C, and the maximum as 42°–44°C.

In more recent studies, the effect of temperature on hypocotyl elongation and seedling emergence has been investigated. Hypocotyl elongation of FORD is inhibited at 25°C while seedlings are normal if germinated at 15°, 20°, and 30°C (Grabe and Metzer, 1969). Twenty-five genotypes were subsequently classified as short, intermediate, or long, based on their ability to emerge at 25°C from a depth of 10 cm of sand, indicating genetic control over the temperature-induced hypocotyl inhibition. Gilman *et al.* (1973) found temperatures between 21° and 28°C inhibited hypocotyl elongation with maximum inhibition at 25°C. Hatfield and Egli (1974) reported the time for hypocotyl elongation to 5 cm decreased as temperature increased between 10° and 32°C with the optimum temperature of 25°–32°C for hypocotyl elongation.

Because of the possibility of poor emergence due to high temperature, Stucky (1976) studied the effect of planting depth and temperature on emergence of several cultivars for double-cropping systems. He concluded that most cultivars can emerge from planting depths of up to 7.5 cm and that high temperature (32°C) accelerated emergence with some reduction in percentage emergence.

2. Root

Earley and Cartter (1945) reported greatest root weights of greenhouse, gravel-grown soybeans occurred at temperatures of either 27° or 32°C, although differences were small for temperature variations between 12° and 37°C. The range of temperature they studied was 7°–37°C.

Temperature affects cation uptake by soybean roots and apparently in a differential manner. Wallace (1957) reported potassium uptake increased but magnesium and calcium uptake decreased as temperature increased from 12° to 32°C. It is reasonable to assume that cation uptake is not maximized for all essential nutrients at the same temperature.

3. Shoot

Leaf development is related to temperature; leaf area increases as temperature increases throughout the range of 18°–30°C (Ciha and Brun, 1975). Brown (1960) proposed a mathematical model, the "soybean development unit," to describe the relationship between soybean development and air temperature. Vegetative development prior to pod filling is closely related to temperature accumulation, while the period between pod filling and maturity is better defined by calendar days rather than temperature accumulation. Major et al. (1975a) evaluated eleven methods of accumulating thermal units as predictors of soybean development. Thermal units with a base temperature subtracted and thermal units using quadratic formulas are better predictors of planting to emergence and emergence to flowering than are calendar days. Calendar days are a better predictor for post flowering development than any of the eleven thermal unit methods.

Iterative regression analysis was used to relate soybean development to temperature and day length (Major et al., 1975b). The cool spring temperatures and long day lengths of northern latitudes delay flowering, but temperature has a greater delaying effect than does day length on early planted soybeans. Both have similar delaying effect when planting occurs after June 1. The time interval between flowering and physiologic maturity increases with cool autumn temperatures and decreases with shorter days. The effect of day length is greater than that of temperature such that the pod-filling interval decreases as planting is delayed.

High air temperature may also have detrimental effects on soybean development. Temperatures throughout the 16°–32°C range increase the percentage of flower and pod shedding (VanSchaik and Probst, 1958). Yield is reduced when July and August temperatures are above average (Runge and Odell, 1960). Seed-filling rate is enhanced by temperatures in the 26°–30°C range compared with the 16°–18°C range (Thomas and Raper, 1976).

Soybeans are susceptible to low temperature injury in early spring though less susceptible to frost than corn, cowpeas, or field beans (Morse et al., 1949). D. R. Hicks (unpublished data) subjected soybean plants at the unifoliolate and third trifoliolate leaf stages to temperatures ranging from −1.1° to −12.2°C for 1 and 2 hours. Unifoliolate plants tolerated lower temperatures without leaf damage. Plants of both stages could tolerate complete leaf killing by a −3.8°C temperature for 2 hours and recover by axillary bud growth from either the unifoliolate or cotyledonary node. Apparently the soybean plant can tolerate lower temperature than thought earlier.

C. Day Length and Light Intensity

Light is important as the energy source in the photosynthetic process. Individual soybean leaves are light saturated at 23,680 lux which is about 20% of full sunlight (Bohning and Burnside, 1956). Leaves from the upper part of a normal canopy of field-grown WAYNE soybeans were recently reported to be light saturated at 107,640 lux and leaves of spaced plants were not light saturated at 161,460 lux (Beuerlein and Pendleton, 1971).

For much of the growing season, only the leaves on the upper periphery of the soybean canopy receive full sunlight. Those below the upper one-half may receive little to no light. Photosynthetic rates of lower leaves of field grown soybeans can be increased by increasing the amount of light which they intercept (Johnston et al., 1969). Although lower leaves apparently are not parasitic to the plant (Shibles and Weber, 1965), it is likely that soybean productivity could be increased by increased lighting of lower leaves.

When light intensity increased from 19,300 to 32,300 lux, leaf area of the terminal leaflet of the third trifoliolate leaf was reduced and stomatal frequenty was increased on both leaflet surfaces (Ciha and Brun, 1975). Kan and Oshima (1952) reported light intensities 50% of normal reduced the number of branches, nodes, and pods and reduced seed yield 60%.

With a 24 hour day length established with incandescent floodlights, days from planting to flowering, plant height and internode number increased as light intensity increased from 2 to 100 lux (Major and Johnson, 1974). Light intensity had no effect on days from flowering to beginning pod fill, number of flowering days, days flowering to maturity, or seed yield.

The main effect of day length on soybean development is that of flowering induction. Soybeans are classified as short day plants because short days initiate the flowering process. The phenomenon has been studied by several researchers and most recently reviewed by Howell (1963).

Soybeans will flower in about 30 days if the day length is short. The length of the dark period is the controlling factor in floral induction. The flowering stimulus originates in the leaf. The leaf that has most recently attained full size is the most effective one in floral induction (Borthwick and Parker, 1938, 1940). The flowering stimulus, called "florigen," is readily translocated both up and down the plant via the phloem. Floral induction can occur at any growth stage after unifoliolate leaf development.

With continued long days, soybeans will remain vegetative almost indefinitely. Flowering is also prevented by a continuous dark or dim light environment. A brief period (1 minute) of light during the dark period will also prevent floral initiation (Hamner and Bonner, 1938).

Floral induction is a reversible reaction. Plants can be induced to flower at certain nodes by exposure to short days and then returned to long days so that subsequent nodes do not produce flowers. When exposed again to short days, flowering occurs at other nodes (Hamner, 1940).

Cultivars adapted to the northern soybean growing areas of the United States will initiate flowering with shorter night intervals (longer days) than cultivars adapted farther south. In the field, the soybean will flower only when the days shorten below a critical value, or conversely when the night length exceeds a critical length. When a cultivar is planted south of its zone of adaptation as a full season cultivar, it flowers at an earlier phenological stage and matures earlier because the critical night length initiating flowering occurs at an earlier calendar date. Conversely, when planted at locations north of its zone of full season adaptation, floral initiation and maturation is delayed because the critical night length occurs at a later calendar date. Since by definition a full season cultivar for any location requires the entire growing season, normal maturation would not occur in an average season before the first killing frost.

Soybean cultivars are adapted to rather narrow ranges of latitude because of responsiveness to photoperiod. Adaptability of a cultivar could be expanded geographically if it were less sensitive to photoperiod. Polson (1972) has identified several genotypes insensitive to day length which flower about 30 days after emergence on photoperiods between 12 and 24 hours.

Photoperiod affects pod production efficiency and seed filling rate. Thomas and Raper (1976) exposed plants, initially grown under long days in a phytotron, to short days at the time of expansion of the primary leaves, when the third trifoliolate was unfolding, or when the sixth trifoliolate was unfolding. The number of pods set per total number of flowers initiated (pod production efficiency) was greater when plants were exposed to short days at the primary leaf stage. However, the number of pods produced per plant was greater when the exposure to inductive short days was delayed until the sixth trifoliolate stage because there were then more nodes per main stem and branches.

Plants were also subjected to various numbers of consecutive short days beginning with expansion of either the primary leaf or the sixth trifoliolate leaf. Continuous exposure to short days after flowering increased dry matter production in pods (includes seeds) at the expense of vegetative tissue. Seed filling rate for plants exposed to 50 consecutive short days after the sixth trifoliolate leaf stage was 13 mg/pod/day compared to 12 and 8 mg/pod/day with 30 and 20 consecutive short days, respectively. Pod growth rates were slower when exposed to short days at the younger morphological age.

D. Soil Aeration

1. Germination

Although the oxygen requirements for germinating soybean seed have not been quantitatively defined, oxygen is necessary for germination (Toole et al., 1956). Respiratory activity requiring an uptake of oxygen increases rapidly with the onset of germination (Howell, 1963). CO_2 is not a necessary component to support germination, but CO_2 concentrations greater than normally found in soils sometimes stimulate germinating soybeans (Grable and Danielson, 1965).

2. Root and Shoot

Ion uptake by roots is generally believed to involve a carrier system which is dependent upon energy derived from respiration (Epstein, 1956). Respiration of soybean roots occurs in the mitochondria which contains the Krebs cycle metabolic system and therefore requires oxygen (Key et al., 1960). When grown in solution culture, soybean roots grow best when the oxygen concentration is about 6 ppm (Gilbert and Shive, 1942). When grown in sand, soybeans can maintain growth processes with oxygen levels as low as 1.5% but top growth is reduced when the root oxygen supply is lower than 1.5% (Hopkins et al., 1950).

E. Soil Condition and Cropping Practices

1. Germination

Emergence from soil by germinating soybean seedlings may be restricted by soil physical condition. Hanks and Thorp (1957) found soybean emergence from soils with low soil crust strengths decreased from 90 to 70% as soil moisture decreased from field capacity to 25% available water. Emergence from soils with high soil crust strengths decreased from 70 to 30% at comparable soil moisture levels.

2. Root, Shoot, and Nodule

Soil compaction increases mechanical resistance to root and nodule growth. The ability of roots to penetrate the soil is negatively related to soil compaction (Zimmerman and Kardos, 1961). Soil compaction by wheel traffic reduces both nodule number and mass (Voorhees et al., 1976). Wheel traffic on both sides of a soybean row causes 30% fewer nodules and 36% less mass than in rows with wheel traffic on only one side. Compaction also reduces nodule mass in the upper 30 cm of the soil profile. Nelson et al. (1975) studied the effect of compaction of a sandy loam soil on the soybean root system and seed yield. Resistance to penetration increased, the root systems were restricted and seed yields were reduced by all compaction treatments compared with no compaction.

Cropping practices may also affect root and nodule development of soybeans. Elkins et al. (1976) studied the effect of soybean cropping frequency on nodule number, mass, and activity rate by inoculating seed with soil sampled from fields with various previous cropping histories. Cropping history did not influence soybean top or root growth if soybeans had previously been grown on the fields. They included soil sampled from fields where soybeans were last grown 11 years previously. Cropping history also did not affect the dominant serogroups present in soybean nodules, although other authors have reported the previous crop has an effect on the serogroups present (Ham et al., 1971a).

Other cropping practices such as planting date and cultivar selection affect the infection nodule serotype. The major portion of the nodules on early planted soybeans was of the 110 and composite of 122 and 125 serogroups while the c2 and c3 serogroups were the major type on late planted soybeans (Caldwell and Weber, 1970). The host genotype causes changes in serogroup populations because of the specificity of certain genotypes for nodulation by specific serogroups (Caldwell and Vest, 1968).

Rhizobia have been found to survive in soils that have not grown a crop of soybeans in 13 years (Lynch and Sears, 1952). Similar survival observations have been reported by others. Ham et al. (1971b) evaluated the effect of several commercial inoculants on nodulation and seed growth of soybeans grown on soils containing naturalized populations of rhizobia. Neither nodulation nor seed yield was enhanced by inoculation.

Naturalized populations of rhizobia in a soil may not contain those strains which are the most efficient in fixing atmospheric nitrogen. Caldwell and Vest (1970) evaluated the effect of 28 strains of R. japonicum on seed yield of five cultivars grown in four environments, of which three were free of R. japonicum. Yields were increased by inoculation with certain rhizobial strains.

When plants are grown in the field with an established rhizobial population, only 5-10% of the nodules may be of the inoculum serotype. Means et al. (1961) reported a single nodule generally contains only one bacterial strain and Johnson and Means (1963) found that rhizobia strains differ in competitiveness in the infecting process. In greenhouse studies, the proportion of nodules formed by the inoculating strain increases as inoculum rate increases (Weaver and Frederick, 1974a). In fields previously cropped to soybeans an inoculum rate of bacteria at least 1000 times that of the naturalized population (per gram soil) is necessary if the inoculating rhizobia are to account for 50% or more of the nodules (Weaver and Frederick, 1974b).

Current data do not support the "inoculation is cheap insurance" concept on soils which have been previously cropped to soybeans. However, the potential exists for increasing soybean seed yield if techniques or methods can be devised for economically introducing and establishing the rhizobial strains which are more efficient in nitrogen fixation than are rhizobial strains which currently exist in the soil.

F. Soil Fertility

1. Root and Nodule

The supply of essential nutrients in the soil affects both root and nodule development. The most notable effect on nodule development and function is brought about by nitrogen. Nodule number, size, and metabolic activity are reduced by increasing soil nitrogen. Weber (1966) reported 33% fewer nodules, 50% reduction in fresh weight, 25% reduction in nodule size, and less nitrogen fixed symbiotically when fertilizer nitrogen was applied.

Raggio *et al.* (1957) reported that nodule formation was inhibited by combined nitrogen (NO_3) in contact with isolated soybean roots, but when nitrogen was fed through the base of isolated roots, nodule formation was not inhibited. Hinson (1975) divided the root system of 17-day-old plants to study the effect of added nitrogen (NH_4NO_3) on nodulation of greenhouse-grown soybeans. He applied nitrogen rates of 0–240 ppm to one-half of the root system and measured nodule number and weight and root and shoot weights after 44 and 57 days. Nodule number and weight were reduced on the root half where nitrogen was added. Nodule number on the root half without added nitrogen was not affected even though the other half of the root system received a high amount of nitrogen. For the root half without added nitrogen, nodule weight was increased by 50% when the other root half received intermediate rates of nitrogen, while nodule weight was either equal to or as much as 40% lower at the highest rate of applied nitrogen. These results indicate a nonlocalized inhibitory effect of combined nitrogen on nodule weight. The nitrogen additions increased shoot weight and total root weight. In field experiments rates of 0, 56, and 112 kg/ha ammonium nitrate applied 1½ or 5 weeks after planting had no effect on seed yield and no detectable effect on nodule development.

In experiments in soil columns of 30 cm depth, Harper and Cooper (1971) placed nitrogen fertilizer either uniformly throughout the column or in the bottom 10 cm to determine whether nodulation inhibition was affected by fertilizer nitrogen placement. Both placement methods of nitrogen caused a reduction in nodule weight and hemoglobin content compared with soybeans grown without added nitrogen. Dispersing the nitrogen throughout the 30-cm column decreased nodule weight and hemoglobin content more than deep placement (bottom 10 cm of column). They concluded combined nitrogen affected nodule development rather than the nodule infection or initiation process.

Even though small amounts of fertilizer nitrogen have been reported to stimulate symbiotic fixation in soybeans (Allos and Bartholomew, 1959), seed yield of soybeans has not been consistently or economically increased by application of fertilizer nitrogen.

2. Shoot

The development of above ground plant parts, including maximum seed yield, is dependent upon the plant obtaining adequate nutrients from the soil. The soil fertility level affects ion uptake by the plant.

Nutrient uptake with time follows the same sigmoid curve as dry matter accumulation. A slow rate of nutrient uptake occurs when plants are young. The rate increases and remains fairly constant until senescence begins when the uptake decreases to zero (Hanway and Weber, 1971b). Average whole plant accumulation rates of 4.5, 0.4, and 1.5 kg/ha/day of N, P, and K, respectively, were reported during full bloom to seed filling of an indeterminate cultivar. Uptake data on determinate cultivars are similar (Henderson and Kamprath, 1970).

Moisture deficiency restricted P uptake in a study by Marais and Wiersma (1975). It is reasonable to assume that available moisture affects the uptake of other essential nutrients also.

Prior to the onset of seed formation, 60% of the N, 55% of the P, and 60% of the K has been absorbed by the plant (Hanway and Weber, 1971b). Since the seed ultimately contains 68% of the total N, 73% of the P, and 56% of the K absorbed, a major source of these elements for seed filling must be the stem and the leaves and petioles which have not fallen from the plant. Developing seeds are a competitive and dominant sink for carbohydrates as well. As a result, the soluble carbohydrate content of roots and stems decreases (Dunphy, 1972), nodule activity declines, root growth stops, and nutrient uptake slows and also ultimately stops as nutrients required for seed filling are retranslocated from vegetative tissues. This phenomenon was termed "self-destructive" by Sinclair and DeWit (1976).

Some combinations of N, P, K, and S fertilizer solutions applied to soybean leaves at late growth stages have caused increased soybean yields (Garcia and Hanway, 1976). Specific rates, application dates, and formulations have not been defined, but the opportunity apparently exists to minimize the "self-destruct" process and increase soybean yields by foliar fertilization.

G. Exogenous Growth Regulators

Since the soybean plant aborts a large percentage of its flowers, increasing pod set has been a major objective of growth regulator research on soybeans. Chemical growth retardants have been used in attempts to decrease vegetative growth during flowering, particularly in indeterminate varieties, so that photoassimilates could be translocated to flowers and pods rather than vegetative growth. The most intensively studied growth regulator for field application is 2,3,5-triiodobenzoic acid (TIBA). Reports (Greer and Anderson, 1965; Hicks et al., 1967; Wax and Pendleton, 1968; Bauer et al., 1969) show that foliar application of TIBA at onset of flowering of field-grown indeterminate soybeans caused less lodging, shorter plants, earlier maturity, more pods per plant, smaller seeds, and yield increases from 0 to 20%. Oplinger (1970) included TIBA in the fertilizer band at planting which did not increase seed yield. Yield increases of 7–10% were obtained with TIBA applied foliarly at three stages of leaf development on BRAGG, a determinate cultivar (Clapp, 1973). Caviness et al. (1968) found no yield change due to TIBA application on some determinate soybeans.

Tanner and Ahmed (1974) reported that TIBA increased yields when growing conditions were good but not when growing conditions were poor. Total dry matter produced and rate of dry matter accumulation were not altered by TIBA, but seed yield and rate of seed dry matter accumulation were greater on TIBA treated plants. Their data support the theory that TIBA acts by slowing the vegetative growth and increasing reproductive growth rather than by improving canopy efficiency.

Seed yield has been slightly increased when a low rate (2.5 g/ha) of 2,4-D was applied at the 3–4 trifoliolate leaf stage followed by TIBA applied at beginning bloom (Johnson and Anderson, 1974).

REFERENCES

Allos, H. F., and Bartholomew, W. V. (1959). *Soil Sci.* **87**, 61–66.
Anderson, C. E. (1961). M.S. Thesis, Purdue University, Lafayette, Indiana.
Bauer, M. E., Sherbeck, T. G., and Ohlrogge, A. J. (1969). *Agron. J.* **61**, 604–606.
Bergersen, F. J. (1958). *J. Gen. Microbiol.* **19**, 312–323.
Bergersen, F. J., and Briggs, M. J. (1958). *J. Gen. Microbiol.* **19**, 482–490.
Bergersen, F. J., and Goodchild, D. J. (1973). *Aust. J. Bio. Sci.* **26**, 741–756.
Bernard, R. L. (1972). *Crop Sci.* **12**, 235–239.

Bernard, R. L., and Singh, B. B. (1969). *Crop Sci.* **9**, 192-197.
Beuerlein, J. E., and Pendleton, J. W. (1971). *Crop Sci.* **11**, 217-219.
Bieberdorf, F. W. (1938). *J. Am. Soc. Agron.* **30**, 375-389.
Bohlool, P. B., and Schmidt, E. L. (1974). *Science* **185**, 269-271.
Bohning, R. H., and Burnside, C. A. (1956). *Am. J. Bot.* **43**, 557-561.
Borst, H. L., and Thatcher, L. W. (1931). *Ohio Agric. Exp. Stn., Bull.* **494**, 1-96.
Borthwick, H. A., and Parker, M. W. (1938). *Bot. Gaz. (Chicago)* **99**, 825-839.
Borthwick, H. A., and Parker, M. W. (1940). *Bot. Gaz. (Chicago)* **101**, 806-817.
Boyer, J. S. (1970). *Plant Physiol.* **46**, 233-235.
Brown, D. M. (1960). *Agron. J.* **52**, 493-496.
Buttery, B. R. (1969). *Can. J. Plant Sci.* **9**, 675-684.
Buttery, B. R. (1970). *Crop Sci.* **10**, 9-13.
Buttery, B. R., and Buzzell, R. I. (1972). *Can. J. Plant Sci.* **52**, 13-20.
Caldwell, B. E., and Vest, G. (1968). *Crop Sci.* **8**, 680-682.
Caldwell, B. E., and Vest, G. (1970). *Crop Sci.* **10**, 19-21.
Caldwell, B. E., and Weber, D. F. (1970). *Agron. J.* **62**, 12-14.
Carlson, J. B. (1969). *J. Minn. Acad. Sci.* **36**, 16-19.
Carlson, J. B. (1973). *In* "Soybeans: Improvement, Production and Uses" (B. E. Caldwell *et al.*, eds.), pp. 17-66. Am. Soc. Agron., Madison, Wisconsin.
Caviness, C. E. (1966). *Crop Sci.* **6**, 211-212.
Caviness, C. E., Thompson, L. F., and Wimpy, T. S. (1968). *Arkansas Farm Res.* **17**, 3.
Chen, A.-P. T., and Phillips, D. A. (1976). *Physiol. Plant.* **38**, 83-88.
Ciha, A. J., and Brun, W. A. (1975). *Crop Sci.* **15**, 309-313.
Clapp, J. G., Jr. (1973). *Agron. J.* **65**, 41-43.
Dart, P. J., and Mercer, F. V. (1964). *Arch. Mikrobiol.* **47**, 344-378.
Decker, R. D., and Postlethwait, S. N. (1960). *Proc. Indiana Acad. Sci.* **70**, 66-73.
Delouche, J. C. (1953). *Proc. Assoc. Off. Seed Anal.* **43**, 117-126.
Delouche, J. C. (1974). *In* "Soybean: Production, Marketing and Use," pp. 46-62. Tennessee Valley Authority, Tennessee.
Dittmer, H. J. (1940). *Soil Conserv.* **6**, 33-34.
Donald, C. M. (1963). *Adv. Agron.* **15**, 1-114.
Doss, B. D., Pearson, R. W., and Rogers, H. T. (1974). *Agron J.* **66**, 297-299.
Dunphy, E. J. (1972). Ph.D. Thesis, Iowa State University, Ames.
Dzikowski, B. (1936). *Pamiet. Panstw. Inst. Nauk. Gospod. Wiejsk.* **16**, No. 253, pp. 69-100.
Earley, E. B., and Cartter, J. L. (1945). *J. Am. Soc. Agron.* **37**, 727-735.
Egli, D. B., and Leggett, J. E. (1973). *Crop Sci.* **13**, 220-222.
Elkins, D. M., Hamilton, G., Chan, G. K. Y., Brigkovich, M. A., and Vandeventer, J. W. (1976). *Agron. J.* **68**, 513-517.
Epstein, E. (1956). *Annu. Rev. Plant Physiol.* **7**, 1-24.
Erickson, E. H. (1975). *Crop Sci.* **15**, 84-86.
Garcia, L. R., and Hanway, J. J. (1976). *Agron. J.* **68**, 653-657.
Gilbert, S. G., and Shive, J. W. (1942). *Soil Sci.* **53**, 143-152.
Gilman, D. F., Fehr, W. R., and Burris, J. W. (1973). *Crop Sci.* **13**, 246-249.
Goodchild, D. J., and Bergersen, F. J. (1966). *J. Bacteriol.* **92**, 204-213.
Grabe, D. F., and Metzer, R. B. (1969). *Crop Sci.* **9**, 331-333.
Grable, A. R., and Danielson, R. E. (1965). *Soil Sci. Soc. Am., Proc.* **29**, 12-18.
Greer, H. A. L., and Anderson, I. C. (1965). *Crop Sci.* **5**, 229-332.
Ham, G. E., Frederick, L. R., and Anderson, I. C. (1971a). *Agron. J.* **63**, 69-72.
Ham, G. E., Cardwell, V. B., and Johnson, H. W. (1971b). *Agron. J.* **63**, 301-303.

Hamblin, J., and Kent, S. P. (1973). *Nature (London), New Biol.* **245**, 28–30.
Hamner, K. C. (1940). *Bot. Gaz. (Chicago)* 101, 658–687.
Hamner, K. C., and Bonner, J. (1938). *Bot. Gaz. (Chicago)* **100**, 338–431.
Hanks, R. J., and Thorp, F. C. (1957). *Soil Sci. Soc. Am., Proc.* **21**, 357.
Hanway, J. J., and Weber, C. R. (1971a). *Agron. J.* **63**, 263–266.
Hanway, J. J., and Weber, C. R. (1971b). *Agron. J.* **63**, 406–408.
Hardman, L. L. (1970). *Diss. Abstr.* **31**, 2401B.
Harper, J. E., and Cooper, R. L. (1971). *Crop Sci.* **11**, 438–440.
Hatfield, J. L., and Egli, D. B. (1974). *Crop Sci.* **14**, 423–429.
Henderson, J. B., and Kamprath, E. J. (1970). *N.C., Agric. Exp. Stn., Tech. Bull.* **197**.
Hicks, D. R., and Pendleton, J. W. (1969). *Crop Sci.* **9**, 435–437.
Hicks, D. R., Pendleton, J. W., and Scott, W. O. (1967). *Crop Sci.* **7**, 397–398.
Hildebrand, A. A. (1959). *Can. J. Bot.* **37**, 927–957.
Hinson, K. (1975). *Agron. J.* **67**, 799–804.
Hopkins, H. T., Specht, A. W., and Hendricks, S. B. (1950). *Plant Physiol.* **25**, 193–209.
Howell, R. W. (1963). *In* "The Soybean" (A. G. Norman, ed.), pp. 75–124. Academic Press, New York.
Hunter, J. R., and Erickson, A. E. (1952). *Agron. J.* **44**, 107–109.
Inouye, C. (1953). *Proc. Crop Sci. Soc. Jpn.* **21**, 276–277.
Johnson, H. W., and Means, V. M. (1963). *Agron. J.* **55**, 269–271.
Johnson, H. W., Borthwick, H. A., and Leffel, R. C. (1960). *Bot. Gaz. (Chicago)* **122**, 77–95.
Johnson, R. R., and Anderson, I. C. (1974). *Crop Sci.* **14**, 381–384.
Johnston, T. J., Pendleton, J. W., Peters, D. B., and Hicks, D. R. (1969). *Crop Sci.* **9**, 577–581.
Kan, M., and Oshima, T. (1952). *Kyushu, Agric. Exp. Stn., Bull.* **10**, 177.
Kato, I., Sakaguchi, S., and Naito, Y. (1955). *Division of Plant Breeding and Cultivation Bulletin* **2**, 159–168. Tokai-Kinki Nat. Agric. Exp. Stn., Ogoso, Isu City, Japan.
Key, J. L., Hanson, J. B., and Bils, R. F.(1960). *Plant Physiol.* **35**, 177–183.
Koller, H. R. (1971). *Crop Sci.* **11**, 400–402.
Koller, H. R., Nyquist, W. E., and Chorush, I. S. (1970). *Crop Sci.* **10**, 407–412.
Li, D., and Hubbell, D. H. (1969). *Can. J. Microbiol.* **15**, 1133–1136.
Liener, J. E. (1976). *Am. Rev. Plant Phys.* **27**, 291–319.
Lillich, T. T., and Elkan, G. H. (1968). *Can. J. Microbiol.* **14**, 617–625.
Ljunggren, H., and Fahraeus, G. (1961). *J. Gen. Microbiol.* **26**, 521–528.
Lockhart, J. A. (1965). *In* "Plant Biochemistry" (J. Bonner and J. E. Varner, eds.), 2nd ed., pp. 826–849. Academic Press, New York.
Lynch, D. L., and Sears, O. H. (1952). *Soil Sci. Soc. Am., Proc.* **16**, 214–216.
Macmillan, J. D., and Cooke, R. C. (1969). *Can. J. Microbiol.* **15**, 643–645.
Major, D. D., and Johnson, D. R. (1974). *Crop Sci.* **14**, 839–841.
Major, D. J., Johnson, D. R., and Luedders, V. D. (1975a). *Crop Sci.* **15**, 172–174.
Major, D. J., Johnson, D. R., Tanner, J. W., and Anderson, I. C. (1975b). *Crop Sci.* **15**, 174–179.
Marais, J. N., and Wiersma, D. (1975). *Agron. J.* **67**, 777–782.
Mayaki, W. C., Teare, I. D., and Stone, L. R. (1976). *Crop Sci.* **16**, 92–94.
Means, V. M., Johnson, H. W., and Erdman, L. W. (1961). *Soil Sci. Soc. Am., Proc.* **25**, 105–108.
Mederski, H. J., Jeffers, D. L., and Peters, D. B. (1973). *In* "Soybeans: Improvement, Production and Uses" (B. E. Caldwell *et al.*, eds.), pp. 239–266. Am. Soc. Agron., Madison, Wisconsin.
Miksche, J. P. (1961). *Agron. J.* **53**, 121–128.

Mitchell, R. L., and Russell, W. J. (1971). *Agron. J.* **63**, 312–316.
Morse, W. J., Cartter, J. L., and Williams, L. F. (1949). *U.S., Dep. Agric., Farmers' Bull.* No. 1520, pp. 1–38.
Nelson, W. E., Rahi, G. S., and Reeves, L. Z. (1975). *Agron. J.* **67**, 769–773.
Nutman, P. S. (1959). *J. Exp. Bot.* **10**, 250–263.
Oplinger, E. S. (1970). Ph.D. Thesis, Purdue University, Lafayette, Indiana.
Polson, D. E. (1972). *Crop Sci.* **12**, 773–776.
Radford, P. J. (1967). *Crop Sci.* **7**, 171–175.
Raggio, M., and Raggio, N. (1962). *Annu. Rev. Plant Physiol.* **13**, 109–128.
Raggio, M., Raggio, N , and Torrey, J. G. (1957). *Am. J. Bot.* **44**, 325–334.
Raper, C. D., Jr., and Barber, S. A. (1970). *Agron. J.* **62**, 581–584.
Reicosky, D. C., Millington, R. J., Klute, A., and Peters, D. B. (1972). *Agron. J.* **64**, 292–297.
Roberts, R. H., and Struckmeyer, B. E. (1946). *Plant Physiol.* **21**, 332–344.
Runge, E. C. A., and Odell, R. T. (1960). *Agron. J.* **52**, 245–247.
Sanders, J. L., and Brown, D. A. (1976). *Agron. J.* **68**, 713–717.
Shaw, R. H., and Laing, D. R. (1966). *In* "Plant Environment and Efficient Water Use" (W. H. Pierre *et al.*, eds.), pp. 73–94. Am. Soc. Agron., Madison, Wisconsin.
Shibles, R. M., and Weber, C. R. (1965). *Crop Sci.* **5**, 575–577.
Shibles, R. M., and Weber, C. R. (1966). *Crop Sci.* **6**, 55–59.
Sinclair, T. R., and DeWit, C. T. (1976). *Agron. J.* **68**, 319–324.
Slatyer, R. O. (1967). "Plant-Water Relationships," p. 366. Academic Press, New York.
Stone, L. R., Teare, I. D., Nickell, C. D., and Mayaki, W. C. (1976). *Agron. J.* **68**, 677–680.
Stucky, D. J. (1976). *Agron. J.* **68**, 291–294.
Sullivan, T. P., and Brun, W. A. (1975). *Crop Sci.* **15**, 319–322.
Sun, C. N. (1955). *Bull. Torrey Bot. Club* **82**, 491–502.
Sun, C. N. (1957). *Bull. Torrey Bot. Club* **84**, 163–174.
Tanner, J. W., and Ahmed, S. (1974). *Crop Sci.* **14**, 371–374.
Thomas, J. F., and Raper, C. D., Jr. (1976). *Crop Sci.* **16**, 667–672.
Toole, E. H., Hendricks, S. B., Borthwick, H. A., and Toole, V. K. (1956). *Annu. Rev. Plant Physiol.* **7**, 299–324.
Ueda, S. (1952). *Proc. Crop Sci. Soc. Jpn.* **21**, 125–126.
VanSchaik, P. H., and Probst, A. H. (1958). *Agron. J.* **50**, 192–197.
Vincent, J. M. (1962). *Proc. Linn. Soc. N.S.W.* **87**, 8–38.
Voorhees, W. B., Carlson, V. A., and Senst, C. G. (1976). *Agron. J.* **68**, 976–979.
Wallace, A. (1957). *Soil Sci.* **83**, 407–411.
Wax, L. M., and Pendleton, J. W. (1968). *Agron. J.* **60**, 425–427.
Weaver, H. L. (1960). *Phytomorphology* **10**, 82–86.
Weaver, R. W., and Frederick, L. R. (1974a). *Agron. J.* **66**, 229–232.
Weaver, R. W., and Frederick, L. R. (1974b). *Agron. J.* **66**, 233–236.
Weber, C. R. (1966). *Agron. J.* **58**, 43–46.
Weber, C. R., Shibles, R. M., and Byth, D. E. (1966). *Agron. J.* **58**, 99–102.
Williams, L. F. (1950). *In* "Soybeans and Soybean Products" (K. L. Markley, ed.), pp. 111–134. (Interscience), Wiley, New York.
Wipf, L. (1939). *Bot. Gaz. (Chicago)* **101**, 51–67.
Wipf, L., and Cooper, D. C. (1940). *Am. J. Bot.* **27**, 821–824.
Woodworth, C. M. (1932). *Ill. Agric. Exp. Stn., Bull.* **384**, 297–404.
Zimmerman, R. P., and Kardos, L. T. (1961). *Soil Sci.* **91**, 280–288.

3

Assimilation

W. A. BRUN

I. Introduction	45
II. Carbon Assimilation	46
A. Photosynthesis	47
B. Photosynthate Distribution	61
III. Nitrogen Assimilation	62
A. Nitrogen Fixation	64
B. Nitrate Reduction	70
IV. Summary	72
References	73

I. INTRODUCTION

The economic end product of soybean production is the seed which comprises about 47% of the total above ground dry weight of the plant at maturity (Hanway and Weber, 1971). The average elemental composition of the above ground dry weight at maturity can be calculated to be approximately 51% O, 38% C, 6% H, 4% N, and lesser amounts of various minerals. The accumulation of these materials from their somewhat random occurrence in the environment into the organized structure of the plant body is referred to as assimilation. It requires not only the availability of the proper substrates in the environment (CO_2, H_2O, N_2, and NO_3^-), but also a great deal of energy in the form of light.

The principal physiological processes responsible for assimilating the plant body from the environment are photosynthesis (assimilating carbon and providing energy) and the two nitrogen assimilatory processes, nitrogen fixation and nitrate reduction. These processes are the subject of this chapter.

II. CARBON ASSIMILATION

The carbon assimilation process or photosynthesis of soybean plants is basically the same as in other C_3 species. C_3 plant species, comprising the great majority of higher plants, are those in which the first detectable products of photosynthesis are 3-carbon compounds (3-phosphoglyceric acid) and in which carbon dioxide enters directly into the classical Calvin cycle of carbon fixation without any preliminary reaction to form 4-carbon acids as in the C_4 plants.

A notable trait of C_3 species is that their respiratory activity is stimulated by light. The increased rate of respiration in light, referred to as photorespiration, is clearly distinguishable from the normal light-insensitive respiratory activity (called dark respiration) which all living cells carry on at all times. Photorespiration is intimately associated with photosynthetic carbon fixation since it has as its substrate the compound phosphoglycollic acid which is synthesized in the chloroplasts by the oxidation of ribulose diphosphate which also serves as the initial carbon acceptor in the classic Calvin cycle of CO_2 fixation.

Photorespiration, unlike dark respiration, is not known to be coupled to the production of useful chemical energy in the cell and for this reason, it is often looked upon as representing an inefficiency in the carbon metabolism of C_3 plants. While it is true that many C_4 plants, which do not display this trait, tend to have higher photosynthetic rates than do many C_3 plants, caution should be exercised in assuming that photorespiration necessarily implies inefficiency.

If there is anything unique about the photosynthetic process in soybeans, it is the use to which the resulting photosynthate is put. Not only must this photosynthate be used for the growth and maintenance of the plant, but the economically important end product, the seed, is remarkably rich in all three of the major food products made by plants (carbohydrates, proteins, and lipids). Soybeans are unique among crop plants in that they classify not only as a good oil crop, but also as an excellent protein crop. The cost in terms of photosynthate for producing both of these concentrated forms of food stuff is very high. Sinclair and de Wit (1975) have calculated that 1 g of lipid costs approximately 3 g of photosynthate (glucose) to produce, while a gram of protein costs 2.5 g, and a gram of typical seed carbohydrate costs 1.2 g of photosynthate to produce. Considering the average chemical composition of soybean seeds to be 42% protein, 31% carbohydrate, 20% lipid, 4% lignin, and 3% ash, the average photosynthate cost for soybean seed is 2.13 g per gram of seed. This compares to a range from 1.33 to 1.45 g of photosynthate per gram of seed for the cereal grains (wheat, oats, rye, corn, popcorn, sorghum, barley, and rice) and a range of from 1.49 to 1.56 for other legumes (chick pea, lima beans, pigeon pea, cowpea, mung bean, pea, and lentil). The photosynthate cost of seed production of soybeans is comparable to that of other oil crops (sunflower, cotton, safflower, flax, hemp, peanut, rape, and sesame) which range from 1.92 to 2.38 g of photosynthate per gram of seed.

Not only is soybean seed costly to produce in terms of photosynthate, but the nitrogen requirements for seed production are much greater than in other crop plants. Sinclair and de Wit (1976) hypothesize that the soybean plant "self-destructs" itself as it approaches maturity because most of its reduced nitrogen compounds are mobilized to the filling seeds.

As discussed below, both of the nitrogen assimilatory systems of soybean plants are costly to operate because they require large quantities of ATP as well as carbon skeletons for amino acid production. The very heavy requirement for reduced nitrogen for seed production thus further accentuates photosynthate requirements during the pod-filling process.

If self-destruction of the photosynthetic portions of the soybean plant is caused by excessive demand for protein during the pod-filling period, then one might well ask whether during evolution of this plant, an optimum relationship, from an economic rather than a survival point of view, has developed between the carbon and nitrogen assimilatory processes.

A. Photosynthesis

1. The Site of Production

The primary site of photosynthesis is in the leaves of the soybean plant. Unlike cereal grains where photosynthetic activity of awns may contribute 50% or more of the photosynthate required by the developing grains, the green soybean pods apparently do not show any net fixation of atmospheric carbon dioxide. Quebedeaux and Challet (1975) have shown that the developing green pods contribute photosynthetically only by reassimilating respiratory carbon dioxide from the seeds, thus conserving carbon which otherwise would be lost to the plant.

Many anatomical, morphological, and physiological traits of soybean leaves relate to their photosynthetic activity. The ability of carbon dioxide to diffuse from the atmosphere to the photosynthetic sites in the chloroplasts within the mesophyll cells normally constitutes a severe limitation to photosynthesis (Brun and Cooper, 1967). The resistance to carbon dioxide diffusion from the atmosphere to the chloroplasts is often partitioned into three in-series components (Gastra, 1959): $R_a^{CO_2}$, associated with the boundary between the leaf epidermis and the atmosphere; $R_s^{CO_2}$, associated with the stomatal openings and intercellular air passages in the leaf; and $R_m^{CO_2}$, associated with the mesophyll cells. The latter diffusion resistance ($R_m^{CO_2}$) is not completely analogous to the other two because it represents not only the physical resistance to carbon dioxide diffusion through the liquid medium of the cell wall, associated membranes, and cytoplasm, but also the limitation to carbon dioxide fixation imposed by the biochemistry of the chloroplasts. Experimentally, $R_m^{CO_2}$ is determined by subtraction of $R_a^{CO_2}$ and $R_s^{CO_2}$ from the total resistance; hence it becomes a depository for many unknowns, recognized or otherwise.

Several leaf traits known to influence photosynthesis can be related directly to the three component diffusion resistances. Leaf pubescence, for example, has been shown to reduce the wind speed 0.5 mm from the leaf surface by 40%, thereby increasing $R_a^{CO_2}$, and reducing gas exchange by the leaf (Wooley, 1964). Also in leaves of near-isogenic lines of CLARK soybeans with densely pubescent, glabrous, and normal leaf surfaces, it has been shown that densely pubescent leaves photosynthesize 23% less than do normal leaves (Ghorashy et al., 1971), presumably because of their larger $R_a^{CO_2}$.

The stomatal resistance to carbon dioxide diffusion ($R_s^{CO_2}$) is dependent upon light, water relations, and stomatal frequency and size. In the absence of light, stomates close causing the $R_s^{CO_2}$ to become overwhelmingly large, thus inhibiting photosynthesis. At low leaf water potentials, the decreased turgor of the guard cells causes stomatal closure, thus increasing the $R_s^{CO_2}$ and inhibiting photosynthesis. Increased stomatal frequency and/or size causes decreased $R_s^{CO_2}$. Ciha and Brun (1975) have examined the stomatal frequency of forty-three soybean genotypes and found it to vary significantly. The mean stomatal frequency on the upper surface of soybean leaves was 130 stomata per mm^2 (ranging from 81 to 174), while on the lower leaf surfaces it was 316 per mm^2 (ranging from 242 to 345). The physiological significance of such variability has apparently not been examined in soybeans, but according to the model of Gastra (1959), it should have a greater impact on transpiration than on photosynthesis, because the stomatal resistance constitutes a relatively larger proportion of the total resistance to water vapor diffusion than to carbon dioxide diffusion. In barley leaves, Misken et al. (1972) have shown that low stomatal frequency lines transpire less than high stomatal frequency lines while maintaining similar photosynthetic rates. Perhaps increased water use efficiency of soybeans might be attainable by breeding for decreased stomatal frequency.

The mesophyll resistance to carbon dioxide diffusion ($R_m^{CO_2}$) is normally the largest of the three component resistances (Jarvis, 1971) and it can be partitioned into a transport resistance and a carboxylation resistance, of which the transport resistance is the greater, at least in cotton (Jones and Slatyer, 1972) and in bean (Chartier et al., 1970). The carboxylation resistance is related to the ribulose diphosphate (RuDP) carboxylase activity of the leaf. A positive correlation has often been found between leaf photosynthetic rate and extractable RuDP carboxylase activity of a number of species including soybeans (Bjorkman, 1968; Bowes et al., 1972; Wareing et al., 1968).

The RuDP carboxylase activity is that enzymatic activity which is responsible for the initial step in the photosynthetic carbon fixation cycle. The RuDP carboxylase enzyme activity is postulated also to have an oxygenase activity by which it can oxidize ribulose diphosphate to phosphoglycolate which is the substrate for photorespiration (Ogren and Bowes, 1971). This enzyme thus has the potential for controlling not only photosynthesis but also photorespiration. The ratio of the carboxylase to oxygenase activity of the purified enzyme from soybean leaves has been found to vary with temperature in a manner similar to the temperature effect on soybean leaf affinities for CO_2 and O_2 (Laing et al., 1974).

Many other leaf traits which influence photosynthetic rates are not so obviously related to the three component diffusion resistances to carbon dioxide.

The sizes and arrangements of the leaflets within the canopy, for example, has been suggested as a trait which may influence canopy photosynthetic rates. Sakamoto and Shaw (1967a) have observed that in field-grown soybean canopies, most of the light is intercepted at the periphery of the canopy and the authors speculate that yield increases could be achieved by selecting for improved light penetration. However, when Hicks and Pendleton (1969) measured the yield from normal and narrow leaflet isolines of both HAROSOY and CLARK soybeans, they were unable to detect any significant differences in yield. Nor could Egli et al. (1970) detect differences in canopy photosynthetic rates of normal and narrow leaflet near isogenic lines of HAROSOY soybeans.

The rate of dry matter production in soybean canopies is linearly related to the percentage interception of solar radiation. Shibles and Weber (1965, 1966) demonstrated this in field-grown HAWKEYE soybeans. They found light interception and dry matter production to increase with leaf area development, reaching a maximum at a leaf area index (LAI: leaf area per unit ground area) of 4. Any further increase in LAI was not associated with decreased rates of dry matter production, indicating that leaf area in excess of that required for full light interception was not a detriment to productivity.

A quantitative evaluation of the effect of leaf orientation on canopy photosynthesis is difficult, but Duncan (1971) has attempted to do it by using a photosynthetic simulation model. Within the limits of the model, he shows that the effect of leaf angle is inseparable from effects of leaf area. For any given leaf area index (LAI), there are arrangements of leaf angles that will give maximum photosynthetic rates. For plant canopies with an LAI of less than 3.0, leaf angle appears to have little practical significance, while with greater LAI leaf angle may be quite meaningful. The optimum leaf arrangement at any given LAI would depend, among other things, on the position of the sun.

The environment prevailing during leaf development clearly influences leaf photosynthetic rates. Beuerlein and Pendleton (1971) suggested that the photosynthetic rate of field-grown WAYNE soybean leaves adapts itself to the light intensity in which they are grown. They found average leaf photosynthetic rates as high as 50 mg CO_2 dm^{-2} $hour^{-1}$ in spaced plants on which branches had been prevented from developing, thus maintaining all the leaves in full sunlight. They attribute the high photosynthetic rate to the light-rich environment of these leaves prior to making the measurements. However, as pointed out by Bowes et al. (1972), the debranching treatment may have been partially responsible for the very high photosynthetic rates.

The saturating light intensity and the maximum photosynthetic rates of the most recently fully expanded trifoliolate leaves of unaltered, field-grown WAYNE soybean plants have been found to be a function of the light intensity received during growth (Bowes et al., 1972). Leaves grown at the highest light intensity (151 klux) had the highest photosynthetic rates (42 mg CO_2 dm^{-2} $hour^{-1}$) and the light intensities required to saturate photosynthesis were approximately those at which the plants had been grown. Soybeans thus appear to develop sufficient, but not excessive photosynthetic capacity to utilize the maximum available light. This light adaptation process is thought (Bowes et al., 1972) to explain the wide range of photosynthetic rates and light saturation values reported for soybeans in the literature.

The relationship between leaf photosynthetic rate and specific leaf weight (leaf weight per unit area) has been investigated by several laboratories with somewhat contradictory results. Dornhoff and Shibles (1970) have shown photosynthesis to be highly correlated with specific leaf weight among twenty soybean varieties, and they suggest that specific leaf weight may be a useful selection criterion for photosynthetic rate in soybeans. The correlation between specific leaf weight and photosynthetic rate, however, may not hold in all environments. Bowes et al. (1972), for example, found no such correlation for plants grown under different light sources, and Mondal et al. (1978) have found specific leaf weight to increase and photosynthetic rate to decrease with decreasing sink demand in soybeans.

The effect of leaf age on photosynthetic rate is difficult to investigate in field plants because older leaves are usually shaded by the younger leaves.

Johnston et al. (1969) have attempted to separate the two effects by providing artificial light to leaves in the bottom, middle, and top of a canopy of field-grown AMSOY and WAYNE soybeans. Without additional lights, the rates of photosynthesis of bottom and middle soybean leaves were 13 and 60%, respectively, of the 20.2 mg CO_2 dm^{-2} $hour^{-1}$ at the top of the canopy. When exposed to full sunlight, bottom and middle leaves photosynthesized 258 and 50% faster, respectively, than when naturally shaded.

Considerable variation exists in the chlorophyll content of soybean leaves in different varieties (Starnes and Dadley, 1965), and a number of chlorophyll-deficient mutants are also known (Bernard and Weiss, 1973). The chlorophyll content of soybean leaves continuously decreases from full expansion to senescence (Wickliff and Aronoff, 1962) but probably does not limit photosynthesis until it becomes very low. Koller and Dilley (1974) have measured photosynthesis in leaves of dark green (normal) and pale green (chlorophyll-deficient mutant) soybeans. When grown at low light intensity (10 klux) the pale green leaves, containing only one-fourth as much chlorophyll, photosynthesized as well as the dark green leaves. Keck et al. (1970) have shown that isolated chloroplasts from the pale green mutant had a maximum rate of electron transport and photophosphorylation about three times greater than did chloroplasts of the normal dark green plant. When grown at high light intensity (30 klux), Koller and Dilley (1974) found that the chlorophyll concentration of the mutant progressively declined to 10% of its value in the 10 klux leaves. Photosynthesis likewise declined as the carbon dioxide saturation concentration declined. At carbon dioxide concentrations below saturation, however, chlorophyll content had no effect on photosynthetic rate.

The relationship between chlorophyll content and photosynthetic rate was investigated in field-grown flowering soybean plants by Buttery and Buzzell (1977). They found that 44% of the variability in the photosynthetic rate of forty-eight cultivars was accounted for by the variability in chlorophyll content, and they suggest that initial screening of progenies in a breeding program for high photosynthetic rate could be done by measuring chlorophyll content.

2. Environmental Effects

Because the photosynthetic rate of soybean leaves depends upon the environment in which they are grown, it is often difficult to interpret the yield implications of photosynthetic rates measured in the growth chamber.

The effects of light intensity on soybean photosynthesis have been studied extensively both in growth chambers and in the field. It is clear that there is a strong interaction between light intensity and carbon dioxide concentration. Brun and Cooper (1967) observed that photosynthetic light saturation of growth chamber-grown HARK and CHIPPEWA 64 occurred at 21.5 klux in normal air (300 ppm CO_2), but at 75.3 klux in carbon dioxide-enriched air (1670 ppm CO_2). Egli et al. (1970), studying canopy photosynthesis in three varieties of field-grown soybeans, also found such an interaction between light intensity and carbon dioxide concentration. At the normally prevailing 300 ppm CO_2, there were significant differences among the varieties but none of them were light saturated in full sunlight.

Soybean plants appear to develop sufficient, but not excessive, photosynthetic capacity to utilize the maximum available light intensity in their environment (Bowes et al., 1972). This environmental adjustment is a result not only of the interaction between photosynthetic light saturation and the light intensity during leaf development (discussed earlier), but also of an interaction between leaf area index development and light intensity. Jeffers and Shibles (1969) have studied this in canopies of three field-grown soybean varieties. They found a strong interaction between leaf area index (LAI) and light intensity. At low light intensity (0.2 langley/minute) there was a critical LAI of 5 to 6, whereas at sunlight intensities (1.2 langley/minute) the critical LAI was above 8. Variability in LAI, however, was obtained experimentally by partial defoliation of the plants and it has since been found in alfalfa that such treatment tends to increase photosynthesis in the remaining leaves (Hodgkinson, 1974).

The effect of light quality on soybean photosynthesis, that is, a photosynthetic action spectrum, has apparently not been determined. Balegh and Biddulph (1970) however, have published such an action spectrum for red kidney beans (*Phaseolus vulgaris* L.) which shows the highest photosynthetic rate to occur in red light, with two peaks (at 670 and 630 nm). Lower peaks in descending order were present in the blue (ca. 437 nm) and in the green part of the spectrum (ca. 500 nm). The action spectrum drops sharply between 680 and 700 nm and between 420 and 400 nm.

Light duration, which has pronounced photomorphogenic effects on soybeans, is not known to have any significant effect on soybean photosynthesis. Brun and Cooper (1967) determined that atmospheric carbon dioxide enrichment to 1670 ppm CO_2 could sustain leaf photosynthetic rates of chamber-grown HARK soybeans at near 60 mg CO_2 dm^{-2} $hour^{-1}$, which is about three times the normal rate, for up to 12 hours without any decline.

The carbon dioxide concentration of the atmosphere around the leaves has been shown by a number of workers to be limiting to photosynthesis. Brun and Cooper (1967) published carbon dioxide response curves for HARK and CHIPPEWA 64 soybeans from a growth chamber. These curves show a lack of carbon dioxide saturation at 1670 ppm carbon dioxide in 75 klux of fluorescent light. At the normally prevailing atmospheric carbon dioxide concentration (300 ppm), photosynthesis shows a linear response to carbon dioxide

3. Assimilation

concentrations at light intensities above 21 klux and some responsiveness at light intensities as low as 5 klux. Egli et al. (1970), using three varieties of field-grown soybeans demonstrated a 72% increase in canopy photosynthetic rate by enriching the surrounding atmosphere with carbon dioxide from 300 to 600 ppm. They concluded that both carbon dioxide concentration and light intensity are limiting photosynthesis. This type of response is, of course, not unique to soybeans and it has been utilized in many greenhouse crops where it is economically feasible to enrich the greenhouse atmosphere with carbon dioxide in order to improve plant yields.

In the open environment of field crops, atmospheric carbon dioxide enrichment is not practical. Harper et al. (1973) have done it in a field of cotton with a LAI of 2.34 to which they added about 224 kg/ha (200 lb/acre) of gaseous carbon dioxide per hour for 10 hours a day. The carbon dioxide concentration at three-fourths plant height was increased to between 450 and 500 ppm which resulted in a 26% increase in daily photosynthesis. The recovery of the added carbon dioxide was from 7 to 33% depending upon light intensities. In soybeans such work has apparently not been reported, although photosynthetic enhancement through atmospheric carbon dioxide fertilization within transparent, partial, or complete enclosures has been used as a research tool to investigate the effects of photosynthesis on other physiological processes (Hardy and Havelka, 1976).

Another atmospheric component, oxygen gas, has a definite inhibitory effect on the photosynthetic rate of C_3 plants including soybeans. The atmospheric oxygen concentration (21% by volume), however, is constant and the absolute quantity present is so large that cultural manipulation of photosynthesis by changing the atmospheric oxygen content is totally out of the question, at least in conventional agricultural systems. As a research tool, however, the knowledge of this effect and its use to manipulate photosynthesis has greatly improved our understanding of the physiology and biochemistry of photosynthesis.

Forrester et al. (1966) reported that the normally prevailing 21% oxygen in the atmosphere resulted in a 30% reduction of apparent photosynthesis in soybean leaves. In the dark, the respiratory rate of these leaves was not affected, but in the light it was found to be strongly stimulated by oxygen, and this stimulation was postulated to be the reason for the decreased apparent photosynthesis. This effect of oxygen on photorespiration and thereby indirectly on the rate of apparent photosynthesis, is now well recognized and documented in C_3 plants.

Oxygen inhibition of photosynthesis consists of two components: (a) a direct inhibition caused by a depletion of Calvin cycle intermediates as oxygen stimulates glycolate synthesis from ribulose diphosphate; and (b) an indirect inhibition caused by the photorespiratory oxidation of glycolate to glycine, serine, and carbon dioxide (Laing *et al.*, 1974). As mentioned earlier, Ogren and Bowes (1971) have postulated that the same enzyme which is responsible for the RuDP carboxylase activity also is responsible for the RuDP oxygenase activity whereby RuDP is oxidized in the presence of molecular oxygen to phosphoglycolate which is the substrate for photorespiration. Laing *et al.* (1974) compared the kinetics of oxygenase and carboxylase activities of the purified enzyme with the kinetics of soybean leaf photosynthesis. They found that with increasing temperature, oxygen inhibition of apparent photosynthesis increased as did the ratio of oxygenase to carboxylase activity of the purified enzyme. This observation is further evidence of a dual function of the enzyme, and indicates that this enzyme has the potential of controlling both apparent photosynthesis and photorespiration in soybean leaves.

The effects of temperature on soybean photosynthetic rates have not been extensively investigated. Jeffers and Shibles (1969) found the temperature response of canopy photosynthesis in three soybean varieties not to be very pronounced, with a poorly defined optimum at 25° to 30°C. There was also some tendency for light saturation to occur at slightly lower irradiation levels at temperatures above the optimum.

The effects of water stress on soybean photosynthesis can be subdivided into at least three categories: (1) effects during leaf growth on traits which subsequently influence photosynthesis; (2) effects of water stress on the stomatal resistance to carbon dioxide diffusion, and (3) effects of water stress on the biochemical events in photosynthesis, that is, effects on $R_m^{CO_2}$.

The indirect photosynthetic effects of water stress through effects on leaf development are related to the turgor requirements for cell expansion. Boyer (1970a) compared the effects of leaf water potential on leaf enlargement and photosynthesis in growth chamber-grown plants of soybean, corn, and sunflower. He found that with increasing water stress, inhibition of leaf enlargement started at the same very mild level of stress (ca. −2 bars) in all three species. At −4 bars, soybean leaf enlargement was reduced by 75% and became nil at −12 bars. Soybean photosynthesis per unit leaf area started to decrease at −11 bars, decreasing rapidly to −18 bars and then much more slowly, reaching 19% of its maximum value when the leaf was desiccated to −40 bars.

Mild water stress, sufficient to inhibit leaf enlargement without directly inhibiting photosynthesis per unit area would thus occur almost daily even in well irrigated field situations where the evaporative demand of the atmosphere often causes plant transpiration to exceed the rate of water uptake by the root system.

The effect of mild water stress during vegetative growth would then be a delay in the development of leaf area. However, since most soybean canopies produce a greater LAI than is required for 95% light interception (Shibles and Weber, 1966), the effects of mild water stress during vegetative growth would be minimal. This is essentially what Doss et al. (1974) found when, from a 3-year study of BRAGG soybeans, it was concluded that an adequate moisture supply was most critical at the pod-filling stage if maximum yields were to be obtained.

The direct effect of water stress on contemporary photosynthesis is largely an effect on stomatal opening (Boyer, 1970b; Mederski et al., 1975). Boyer (1970b) examined the rates of photosynthesis and transpiration in corn and soybean plants undergoing gradually increasing water stress. He found that in both species the primary factor limiting photosynthesis is the diffusive resistance of the stomates to CO_2. Soybeans appeared to be somewhat more resistant to water stress because stomatal closure started at -11 bars, compared to -3.5 bars in corn. Between -11 and -16 bars, soybean leaf photosynthetic behavior was attributed solely to differences in stomatal behavior. At leaf water potentials below -16 bars, factors other than stomatal opening appeared to be involved, but their effect was much smaller. Mederski et al. (1975) using an entirely different technique arrived at essentially the same conclusion in that they found no evidence for other than a stomatal effect on soybean photosynthesis under water stress.

Direct inhibitory effects of water stress on leaf photosynthesis through dehydration of the cytoplasm and/or chloroplasts have been postulated in a number of plants. Boyer (1965) found decreasing photosynthetic rates with decreasing leaf water potentials of cotton leaves grown in such a way that stomatal resistances did not change. This effect was attributed to changes in the mesophyll cells themselves rather than to changes in stomatal aperture. Plaut (1971) observed a partial inhibition of photosynthetic carbon dioxide fixation by isolated chloroplasts from spinach leaves, when exposed to increasing osmotic potentials of sorbitol. This effect is attributed to decreased activities of enzymes associated with the regeneration of the CO_2 acceptor, ribulose-1,5-diphosphate.

This type of work, however, has apparently not been done on soybeans, and considerable evidence (discussed above) seems to indicate that the overwhelming effect of water stress on soybean photosynthesis is on stomatal closure which limits the availability of carbon dioxide to the leaf.

3. Diurnal and Seasonal Patterns

The diurnal as well as the seasonal patterns of canopy photosynthesis by field-grown soybeans represent the integration of all of the environmental and developmental effects on the process. Any attempt to relate soybean productivity, either total biological productivity or the economic productivity (seed yield and quality), to photosynthesis must take into account the diurnal and seasonal patterns of photosynthesis as well as the patterns of dark respiration exhibited by the plants. The latter is a much neglected area compared to the former which has been well documented for a number of environments.

The interest in diurnal photosynthetic rates relates to two questions: When is it most appropriate to sample photosynthesis, and to what extent, if any, is photosynthesis inhibited by an accumulation of its end products?

A diurnal trend in photosynthesis of young growth chamber-grown LEE soybeans maintained in an environment similar to that prevailing during plant development was observed by Pallas (1973). It showed a very low (6–8 mg CO_2 dm^{-2} $hour^{-1}$) but constant photosynthetic rate for 13 hours of the 16-hour day. For the last 3 hours photosynthesis dropped off to a final value of 3 mg CO_2 dm^{-2} $hour^{-1}$. The author suggests that this decline may be due to an "endogenous rhythmic change" in diffusive resistance and/or biochemical activity of the leaves.

Diurnal photosynthesis curves for soybeans have also been recorded by Upmeyer and Koller (1973). This work was on the unifoliolate leaves of 11-day-old BEESON soybeans grown in nutrient solution culture in a growth chamber. Photosynthetic rates and carbohydrate levels of the unifoliolate leaves were determined during a 16-hour, constant environment day similar to the environment in which the plants were raised. The results show nearly constant photosynthetic rates (between 31 and 34 mg CO_2 dm^{-2} $hour^{-1}$) for most of the day followed by a slight decline to about 30 mg CO_2 dm^{-2} $hour^{-1}$ during the last 2 hours of a 16-hour day. The authors suggest that the slight photosynthetic decrease at the end of the day is caused by an increase in soluble carbohydrate level. Nafziger and Koller (1976), however, saw no feedback inhibition of photosynthesis by accumulated soluble sugars in soybean leaves pretreated with high carbon dioxide levels. They did find an inhibition of photosynthesis by the treatment but it appeared to be associated with a starch buildup in the chloroplasts which may impede intracellular CO_2 transport.

If end product inhibition of photosynthesis occurs, it should be particularly severe in leaves made to photosynthesize at greater rates than normal. Brun and Cooper (1967) have measured photosynthesis in leaves of chamber-grown HARK soybeans in which photosynthesis was stimulated to 75 mg CO_2 dm^{-2} $hour^{-1}$ by carbon dioxide enrichment to 1670 ppm CO_2. Under these conditions of abnormally high photosynthetic rates, no detectable decrease in photosynthetic rate was found during a 12-hour day.

Under field conditions the environment is, of course, constantly changing so one cannot extrapolate diurnal rates obtained in the growth chamber to field conditions. Sakamoto and Shaw (1967b) working with field-grown HAWKEYE soybeans in 76 cm rows, measured diurnal photosynthetic rates of

3. Assimilation

0.7 m² of canopy enclosed in a plastic assimilation chamber. A typical curve for a clear day shows photosynthesis to increase during the morning hours with light intensity. From 11 A.M. until 2 P.M. photosynthetic rate was fairly constant and then it diminished in the afternoon with declining light intensity. The midday plateau is interpreted by the authors as being due to light saturation, which they report at about 6000 ft-c (i.e., ca. 60% of full sunlight). They also suggest that the plateau may have been related to moisture stress, although they tend to discount this because there was water condensation on the inside walls of the assimilation chamber. Without data on leaf temperature or leaf water potentials, it may, however, have been unwise to discount the possibility of stress-induced stomatal closure as being responsible for the midday plateau in the observed photosynthetic rate.

The seasonal pattern of photosynthesis in soybean communities is also of considerable interest, particularly with regard to the relative contribution of such photosynthesis at different growth stages to the ultimate yield obtained. Quantitative field data on canopy photosynthetic rates of soybeans over the entire season are apparently lacking. Sakamoto and Shaw (1967b) show such data from flowering until maturity for HAWKEYE soybeans. These data show a maximum photosynthetic rate to occur at initial flowering, dropping slightly to a constant rate through pod formation and early bean filling, but then dropping sharply during the latter part of bean filling as the LAI rapidly decreased at senescence.

Leaf photosynthesis of twenty soybean varieties was measured from anthesis to maturity by Dornhoff and Shibles (1970). Photosynthesis increased after the first week in August by which time eighteen of the twenty varieties were in the filling stage. They speculate that this increase in photosynthesis may be due to an increased "sink demand" from the filling seeds.

The question of the relative contribution of photosynthesis at various stages of growth to the ultimate yield of soybeans has been examined by Hardman and Brun (1971) using spaced, field-grown HARK soybeans in which photosynthesis was stimulated by atmospheric CO_2 enrichment (to 1200 ppm CO_2) for 5-week time periods during vegetative, flowering, and pod-filling stages of development. Carbon dioxide enrichment during vegetative growth had no discernible effect on vegetative or reproductive parameters measured at maturity. Carbon dioxide enrichment during flowering caused increased node numbers, leaf and stem dry weights, and pod numbers at maturity, but seed size decreased so that seed yield was not affected. Carbon dioxide enrichment during pod filling had no effect on vegetative parameters, but caused a marked increase in seed yield, associated with a slight increase in pod numbers.

The often expressed disappointment in the apparent poor correlation between the photosynthetic rates of different soybean genotypes and their yield (Curtis *et al.*, 1969; Dornhoff and Shibles, 1970) may be due to the fact that varietal differences in photosynthetic rate are usually investigated in seedling or young vegetative plants. The data of Hardman and Brun (1971) would indicate that photosynthesis at this time has a minimal effect on the final yield.

4. Genetic Differences

Early attempts to look for genetic differences in photosynthetic rates of soybeans were often done on chamber-grown young plants (Dreger *et al.*, 1969; Curtis *et al.*, 1969), and may thus not have been relevant to a search for a selection index in a yield improvement program. Curtis *et al.* (1969) concluded that varietal yield differences were not caused by the differences in photosynthetic rates which they detected among thirty-six varieties of soybeans grown in a controlled environment chamber.

Dornhoff and Shibles (1970), using twenty field-grown, containerized soybean genotypes, measured leaf photosynthetic rates at carbon dioxide concentrations ranging from 100 to 400 ppm carbon dioxide. They found significant differences in photosynthetic rates among genotypes at all carbon dioxide concentrations. Seed yields were not measured but the authors state that there were several examples of a poor correlation between the observed photosynthetic rates and the known yield potentials of the varieties. They emphasize that photosynthesis is only one factor affecting yield.

5. Internal Control Mechanisms

Several investigations indicate that soybean photosynthetic rates are subject to internal regulatory mechanisms which control leaf photosynthetic rates as the plant develops, especially during reproductive growth (Sakamoto and Shaw, 1967b; Dornhoff and Shibles, 1970). As indicated earlier, the photosynthetic rate during pod filling is particularly important for the final yield of soybeans (Hardman and Brun, 1971) so there is considerable interest in the mechanisms underlying changes in photosynthesis at this time. There appear to be two effects on photosynthesis during pod filling: (1) the presence of filling pods stimulates photosynthesis in the leaves, and (2) the leaves show a gradual senescence progressing from the lower to the upper leaves (Mondal *et al.*, 1978). Whether there is a common mechanism underlying these two changes in photosynthesis is not at all clear, nor is there a good consensus on what causes either one. The fact that the two changes are in opposite directions may indicate a fundamental difference in their underlying mechanisms.

The stimulatory effect of filling pods was observed by Dornhoff and Shibles (1970) who found photosynthetic rates of twenty soybean genotypes to increase 23.4% after pod filling started. They ascribe this to a "sink demand" from the developing pods, but find it difficult to relate it to an increase in specific leaf weight which they also observed at this time.

A parallel observation was made by Lawn and Brun (1974) who found that partially depodding CHIPPEWA 64 and CLAY soybeans caused a 30 and 38% decrease in canopy photosynthetic rates of the two varieties, respectively, 9 days after treatment. Mondal *et al.* (1978) have also observed a decreased leaf photosynthetic rate of HODGSON soybeans 32 hours after removing all reproductive and vegetative sinks from the plants. The absolute magnitude of this effect remained constant from mid-bloom through maturity while its relative effect increased due to the gradual senescence of photosynthetic rate in the control plants.

Similar observations showing that fruit removal decreases leaf photosynthetic rate have been made in other species (Hansen, 1971; Lenz, 1974; Loveys and Kriedeman, 1974; Flinn, 1974). Several mechanisms have been proposed for such observations: (1) end product inhibition of photosynthesis by soluble carbohydrates; (2) excessive starch grain formation within the chloroplasts; and (3) endogenous hormonal signals operating between the photosynthetic sinks and sources.

The hypothesis that leaf photosynthetic rates are inhibited by an accumulation of soluble carbohydrates (reviewed by Neales and Incoll, 1968) appears not to be valid for soybeans. Thorne and Koller (1974), as well as Nafziger and Koller (1976) and Mondal *et al.* (1978), all fail to find a positive correlation between leaf photosynthetic rate and leaf soluble carbohydrate concentrations in soybean plants in which the source–sink relations were manipulated in various ways.

End product inhibition of photosynthesis has also been ascribed to the formation of insoluble starch grains within the chloroplasts. In soybeans, Nafziger and Koller (1976) have found a strong negative correlation between net photosynthetic rate and starch content of leaves of young soybean plants pretreated with 50, 300 or 2000 ppm carbon dioxide for 12.5 hours prior to photosynthetic measurements at 300 ppm carbon dioxide. They found that the reduced photosynthetic rate in leaves with high starch concentrations was due to an increased mesophyll resistance to carbon dioxide diffusion ($R_m^{CO_2}$). As discussed above, $R_m^{CO_2}$ contains both a biochemical and a physical component. Since the carbon dioxide compensation concentration was not influenced by starch accumulation, the authors suggest that the effect of chloroplast starch is primarily an effect on the length of the diffusion pathway in the starch-filled chloroplasts.

Leaf photosynthetic rates and starch concentrations have been examined in reproductive soybean plants with altered source–sink relations and as a function of senescence (Mondal *et al.*, 1978). It was found that desinking the plants (pod removal) caused some starch accumulation in the leaves but this correlated poorly with photosynthetic rates, especially during leaf senescence.

Endogenous hormones have also been implicated in controlling leaf photosynthesis, although apparently none of this work has been done on soybeans. Hormonal regulation of photosynthesis may involve stomatal responses, chloroplast responses, or translocation responses. Loveys and Kriedemann (1974) found that the decreased stomatal diffusion resistance ($R_s^{CO_2}$) in leaves of fruiting grape vines was associated with decreased levels of both abscisic acid (ABA) and phaseic acid (PA). ABA is well known to be associated with stomatal closure and consequent photosynthetic inhibition in response to water stress. PA is a metabolite of ABA which has been shown to have a strong inhibitory effect on photosynthesis in grape leaves (Kriedemann et al., 1975), possibly through an effect on photosynthetic electron flow.

Whether the stimulation of soybean photosynthesis by the presence of pods is related to ABA concentrations is not known, nor is it known whether PA has any effect on soybean leaf photosynthesis, although it is present in soybean leaves (M. L. Brenner, personal communication).

Other endogenous hormones known or thought to be involved in regulation of photosynthesis in legumes, but not specifically in the soybean, include auxins, cytokinins, and gibberellins.

Turner and Bidwell (1965) found that spraying bean leaves (*Phaseolus*) with IAA (indole acetic acid) solutions stimulated photosynthesis and hypothesized that the stimulatory effect of lateral bud break on photosynthesis was associated with an endogenous IAA signal from these buds. Later, Tamas et al. (1974) reported that IAA stimulated photosynthesis in oat and pea leaves by increasing the coupling between electron transport and photophosphorylation. They believe this to be part of a hormonal control mechanism ensuring an increased supply of photosynthates to actively growing sinks. Other endogenous hormones have also been reported to influence photosynthesis. Treharne et al. (1970) have shown increased rates of photosynthesis in leaves of maize, bean, and clover sprayed with solutions of gibberellic acid (GA_3) and kinetin. This effect was associated with an increase in the RuDP carboxylase activity of the leaves which appeared to be related to enzyme activation rather than enzyme synthesis.

The onset of leaf senescence in soybeans as the pods mature is well known to anyone observing the plants growing in the field. It is usually considered that the functional senescence (reduced photosynthesis) of the canopy which also takes place at this time (Sakamoto and Shaw, 1967b) is related to the observable decrease in chlorophyll content of the leaves. As discussed earlier, Sinclair and de Wit (1976) have postulated that soybean leaf senescence is caused by nutrient mobilization, especially of proteins, toward the developing pods with their high protein seeds. However, the findings of Mondal et al. (1978) that deterioration of leaf photosynthetic rate occurs whether or not the leaves turn yellow, and irrespective of whether pods are present on the plant, seems to suggest quite a different mechanism.

B. Photosynthate Distribution

Perhaps equally important as photosynthate production is its distribution to the various metabolic sinks in the plant. Not only is the distribution of carbohydrate important, but, as discussed earlier, because of the unique chemical composition of soybean seeds, the distribution of reduced nitrogen compounds is also of great importance. The latter is, of course, intimately associated with the former because both symbiotic nitrogen fixation in the nodules and nitrate reduction in the leaves is dependent upon photosynthesis for both energy and carbon skeletons.

Photosynthate transport takes place through the phloem tissue (sieve tubes composed of sieve tube elements) of the vascular strands. These vascular strands (referred to as veins in the leaf, vascular bundles in the stem, and as the stele in the root) comprise a highly branched interconnecting system extending from root tips to leaf tips.

As pointed out by Evans (1976) there is some uncertainty about the details of the mechanisms of phloem translocation, but it seems clear that transport over this system follows the path of least resistance so that it occurs primarily by the most direct route, with individual sinks being supplied by the nearest sources. In addition, there appears to be a competitive advantage of sink size such that at equal distance from the source, larger sinks tend to be favored disproportionately to their size (Cook and Evans, 1976).

In young soybeans, as in other plants, photosynthate translocation out of the leaves occurs almost entirely as sucrose (Clauss et al., 1964), although in another legume, the lupine, it has been found that in fruiting plants an increasingly large proportion of the carbon arriving in the pods arrives as amino acids, primarily asparagine. The amide nitrogen of asparagine is then used for the synthesis of other amino acid in the pods.

The patterns of assimilate distribution in soybean plants have been investigated primarily by letting a designated source leaf photosynthetically assimilate $^{14}CO_2$ for a short time and then sampling plant parts at various times afterward for radioactivity. There appear to be two distinct patterns of distribution of labeled photosynthate in soybean plants. In young vegetative plants, photosynthates are distributed to the nearest meristematic regions (Thaine et al., 1959; Thrower, 1962). This means that as leaves develop they first act as sinks for their own photosynthate and for that of more mature leaves further down the stem. Developing leaves change from importers to exporters of photosynthate when they are about 50-60% fully expanded. With further ontogeny the leaf becomes at first a major source for the apical sink and then gradually a source for the root system.

A quantitative balance sheet for the movement of photosynthetically fixed carbon is apparently not available for soybeans, but Minchin and Pate (1973) and Pate (1976) have published such for vegetative pea and lupin plants, respectively. They find that 74 and 71% of the fixed carbon is utilized in the root systems of the two plants, respectively. Of this, 47 and 40%, respectively, is respired to the soil, and 15 and 9% recycled to the stem as amino compounds produced by the nodules.

As the plants set fruit and the pods start to develop, the distribution pattern for photosynthate shifts in favor of the developing pods. Again complete balance sheets are not available for this growth period, but Hume and Criswell (1973) have shown that after rapid seed development begins, roots and nodules accumulate very little photosynthate. During the pod-filling period nitrogen fixation by nodules declines very sharply, which has been attributed to competition from the developing pods for the available photosynthate (Lawn and Brun, 1974).

Because photosynthate moves through the phloem it has been of some interest to determine to what extent, if any, the amount of phloem tissue in a leaf and its connecting vascular strands limit assimilate movement and thus possibly the photosynthetic rate. Gallaher et al. (1975) have compared two C_4 and C_3 species with respect to their photosynthate translocation capabilities and vascularization. They find that the two C_4 species had a greater translocation capacity and a greater cross-sectional area of phloem in their leaves and petioles. The contention, however, that leaf photosynthetic rates might be limited by the transport capacity of the phloem serving the leaf implies control of photosynthesis by end product inhibition. As discussed above, such end product inhibition of photosynthesis appears not to occur in soybean leaves, and thus it would appear that transport capacity would perhaps not be expected to limit photosynthesis of soybeans.

Sanders et al. (1977) have recently examined the effect of experimentally altered rates of photosynthesis on the cross-sectional phloem area in BRAGG soybeans raised in the greenhouse and growth chambers. They found no change in petiole phloem cross-sectional area in response to changes in leaf area or photosynthetic rate, and therefore, conclude that photosynthate supply does not affect phloem development.

III. NITROGEN ASSIMILATION

As pointed out earlier, the combination of high protein and high oil content of soybean seed makes the assimilation of nitrogen critical to the attainment of maximum yields. The soybean plant, like most other legumes and a few nonlegumes, has two different systems of nitrogen assimilation. Like other plants they can absorb fixed or chemically combined forms of nitrogen from the soil through their roots. This nitrogen, usually in the form of nitrate ions (NO_3^-) is then translocated through the xylem to the leaves, where energy derived from photosynthesis (NADH) is utilized to reduce the NO_3^- (by the enzyme system nitrate reductase) to amino nitrogen which is then incorporated into various amino acids. These amino acids may be assembled into proteins in the leaves or they may be translocated to other sinks and there assembled into protein.

In addition to this system, soybeans and other nodulated legumes can assimilate molecular nitrogen gas (N_2) which comprises 80% of our atmosphere, in the root nodules. These nodules contain symbiotic associations between the root cells of the soybean host and a normally saprophytic soil bacterium, *Rhizobium japonicum*. Molecular nitrogen diffusing from the soil atmosphere into the nodule is reduced to amino nitrogen utilizing energy derived from the respiration of photosynthates translocated to the nodules from the leaves. The amino nitrogen is then transported in the xylem, primarily as asparagine, to the leaves where it is converted to various amino acids and then moved via the phloem to whatever nitrogen sinks there are at that stage of development in the plant.

This ability of nodulated legumes to utilize the abundant but otherwise essentially unavailable molecular nitrogen in the environment has undoubtedly been of evolutionary and survival value to legumes. This ability in a crop plant such as the soybean has recently taken on added significance in light of the scarcity and high cost of energy resources which are spent for the production of commercial nitrogen fertilizer.

The relative metabolic cost to the plant of assimilating nitrogen by the two systems has not been compared in soybeans, but in other legumes it appears to be very similar. Gibson (1965, 1974) examined the question in subterranean clover by a growth analysis technique. He found that during a period of active nitrogen fixation, the carbohydrate requirements per milligram of nitrogen fixed were a statistically insignificant 0.7 mg higher than were the requirements for assimilation of the same amount of nitrogen from NH_4NO_3. During nodule development, when N_2 fixation was minimal, there was an apparent decreased growth rate which was attributed to the carbohydrate requirements for nodule initiation.

There have been many attempts to assess the relative contribution of the two nitrogen assimilatory systems to the nitrogen nutrition and yield potential of soybeans (Harper, 1974; Weber, 1966a,b). Interpretation of such studies is often difficult because, as discussed elsewhere, the N_2 fixation process is depressed by combined forms of nitrogen in the soil so that the higher the soil nitrogen level, the less symbiotic N_2 fixation occurs in the plants. In order to better compare relative contributions of the two systems, genetic isolines (or near isolines) differing only in their ability to nodulate, and thus to fix N_2 symbiotically, are often used. Weber (1966a,b) examined agronomic traits and chemical composition of such a pair of isolines under varying environmental and soil conditions. As expected, plant growth and yield of the nonnodulating isoline depended very strongly on the level of available fertilizer nitrogen in the soil. The nodulated isoline, in contrast, responded much less to applied fertilizer nitrogen. The amount of N_2 fixed by the nodulating isoline depended strongly on the amounts of fertilizer nitrogen in the soil, and upon soil moisture conditions. The amount of N_2 fixed decreased with increases in fertilizer nitrogen. Nitrogen fixation ranged from 1 to 160 kg of nitrogen per ha which supplied from 1 to 74% of the plant's nitrogen requirements.

The relative contribution at various times during the growing season of the two systems to the plant's nitrogen nutrition and to the final seed yield has been studied by Harper (1974), using nodulating and nonnodulating soybean isolines grown in soil and hydroponically in outdoor gravel cultures. Seasonal profiles of the two processes indicated that the nitrate reductase system of both isolines gradually increased as leaf area per plant increased. Nitrate reductase peaked at full bloom and declined rapidly thereafter. The N_2 fixing system, on the other hand, got a much slower start and then peaked about 3 weeks later than did the nitrate reductase activity. Both nitrogen assimilation systems appeared necessary in order to obtain maximum yields. Some soil nitrate was necessary for seedling establishment, but too high a nitrate level inhibited the N_2 fixation process upon which the plant depended almost exclusively during the pod-filling period.

A. Nitrogen Fixation

1. The Site of Fixation

The site of N_2 fixation is in nodules. Nodules develop following infection of the root by *Rhizobium japonicum* bacteria from the soil as described in Chapter 2.

Young nodules develop the ability to fix N_2 when they develop an internal pink color due to the presence of leghemoglobin. Leghemoglobin is a soluble red hemoprotein pigment which is capable of being reversibly oxygenated. Most recent evidence for the function of leghemoglobin is that it acts as an oxygen carrier which, by a process known as facilitated diffusion, transports oxygen through the nodular tissue to the respiratory sites, thus providing the necessary oxygen for oxidative phosphorylation to make the required ATP, without damaging the highly oxygen-sensitive nitrogenase enzyme system of the bacteroids which is responsible for the fixation of N_2.

2. The Measurement of N_2 Fixation

The rate of N_2 fixation can be measured or estimated in several ways. In some studies the obvious pink color of the nodule interior is used as a simple nonquantitative criterion of nodule activity.

Quantitative estimates of the rate of N_2 fixation can be made by various techniques. Periodic determinations of the Kjeldahl nitrogen content of the plants will give an estimation of the overall nitrogen assimilation, but fails to specifically estimate N_2 fixation unless the plants are growing in a medium without any combined nitrogen.

Kjeldahl analyses have also been used in conjunction with nodulating and nonnodulating soybean isolines (Weber, 1966b). By subtracting the total nitrogen content of the nonnodulating line at the end of the season from that of the nodulating line, an estimate of the symbiotically fixed nitrogen was obtained. This technique, however, assumes that the two lines remove the same amount of combined nitrogen from the soil, which may not be the case particularly if soil nitrogen is low so that the nonnodulating line grows slower than the nodulating line. Conversely, at higher soil nitrogen levels, the growth and nitrate absorption by the two lines may be more nearly equal, but the symbiotic fixation will be suppressed by the high soil nitrogen.

A stable isotope of nitrogen, ^{15}N, has also been used to estimate symbiotic N_2 fixation. ^{15}N, however, is expensive as is the instrumentation necessary to detect it, the mass spectrograph. Two different ^{15}N procedures have been used. Burris et al. (1943) exposed a nodulated root system to ^{15}N-enriched nitrogen gas and later measured the $^{15}NH_3$ content of the plant tissue. While this technique provides a direct sensitive measure of nitrogen fixation, it suffers from the drawback that the root system must be placed in a gas-tight container while exposed to the expensive ^{15}N-enriched nitrogen gas.

A somewhat different approach is to administer the ^{15}N as ^{15}N-labeled fertilizer to the soil in which plants are growing. This causes the plant to assimilate some ^{15}N by the nitrate reductase system and thus a certain ratio of ^{15}N to ^{14}N can be found in the tissues of the plant. If the plant is also fixing nitrogen symbiotically, this ratio of ^{15}N to ^{14}N in the plant will be somewhat less than it was in the soil. By comparing the $^{15}N:^{14}N$ ratio of the plant tissue to this ratio of the soil nitrogen it is possible to estimate the rate of symbiotic N_2 fixation in the plant.

The ^{15}N tracer technique is more adaptable to field studies than is the $^{15}N_2$ gas technique, but it is still expensive and suitable only for estimating N_2 fixation over relatively long periods of time. It assumes that the availability of ^{15}N-labeled fertilizer is exactly the same as that of the indigenous soil nitrogen.

The most expeditious and commonly used estimate of N_2 fixation is derived from the acetylene reduction assay. This assay is based upon the independent observations of Dilworth (1966) and Schollhorn and Burris (1966) that nitrogenase preparations can reduce not only the triple bond between atoms in nitrogen gas, but also the triple bond between carbon atoms in the acetylene molecule. The product of the acetylene reduction is ethylene gas. Hardy et al. (1968) have described the methodology and application of an assay for nitrogenase activity of legume nodules, based upon their ability to reduce acetylene to ethylene. The assay which utilizes a gas chromatograph with a flame ionization detector, is simple, rapid, and sensitive. The assay consists of incubating legume nodules at room temperature in an atmosphere containing 10% acetylene. After a given incubation time an aliquot of the gas is withdrawn and injected onto the gas chromatograph for the quantitative estimation of ethylene.

The acetylene reduction assay has been applied to many different plant systems, including detached nodules, intact root systems, soil cores containing nodulated roots, and rooted plants. It has also been modified to involve periodic nondestructive determinations on intact plants with their roots exposed to a gas stream containing small amounts of acetylene (Pirl, 1976).

The acetylene reduction assay is indicative of the nitrogen fixation potential of the nodules examined. It is most useful for determining relative treatment effects rather than for estimating absolute rates of N_2 fixation. The reason is that the appropriate ratio between acetylene reduction and N_2 fixation is somewhat uncertain. On theoretical grounds this ratio should be 3 since it takes six electrons to reduce each mole of N_2, but only two electrons to reduce a mole of acetylene to ethylene. However, this ratio may be unrealistically small since it has been shown that with most legumes, including soybeans, a significant portion of the electron flow in the nitrogenase system is to H^+ rather than N_2 (Schubert and Evans, 1976). In the presence of acetylene this reduction of H^+ to H_2 gas does not occur and therefore the acetylene reduction assay tends to overestimate the rate of N_2 fixation. In order to quantify this overestimation, Schubert and Evans (1976) have introduced the concept of the relative efficiency of N_2 fixation which they define as:

$$RE = 1 - \left[\frac{\text{rate of } H_2 \text{ evolution in air}}{\text{rate of acetylene reduction}}\right]$$

Thus the relative efficiency of N_2 fixation expresses the percentage of the total electron flow through the nitrogenase enzyme system which reduces N_2 to NH_3. Schubert and Evans (1976) found that with most legumes, including soybeans, the relative efficiencies of N_2 fixation were only 40-60%. The effects of environment and the physiological condition of the host plant on the relative efficiency of N_2 fixation are unknown at this time except for a recent paper by Bethlenfalvay and Phillips (1977) showing RE of N_2 fixation by pea plants to increase from 40 to 70% between the onset of flowering and the period of rapid pod filling, thus suggesting that the RE may be subject to control by source-sink relationships.

3. Environmental Effects

The effects of various environmental factors on N_2 fixation has been the subject of a great many investigations.

As mentioned above, oxygen is essential for symbiotic N_2 fixation to occur, because of its function in the oxidative phosphorylation of respiratory substrates which supplies the required ATP to drive the reaction. The unique role of leghemoglobin in providing this oxygen at the respiratory sites without inhibiting the oxygen-sensitive nitrogenase enzyme system has also been mentioned. For this reason any cultural or other manipulation of the environment which alters the availability of oxygen to the nodules has a pronounced effect on N_2 fixation by such nodules (Bergersen, 1970). The effects of oxygen concentration on nitrogen fixation appears to depend upon plant age. Criswell et al. (1977) found the optimum oxygen concentration for acetylene reduction by excised KENT soybean roots to decline from 41 to 26% oxygen during a 30-day period in August while the pods were enlarging. The nodules on intact KENT soybean roots, however, were somewhat less sensitive to oxygen. In natural soil environments, oxygen concentrations are similar to or less than the 21% in the above ground atmosphere, but vary considerably with soil moisture, texture, and microbial activity. Criswell et al. (1976) have, however, observed that soybean root nodules can adapt fairly rapidly to such low oxygen levels. Exposure of intact nodules to 6% oxygen, for example, caused an initial decrease in activity of 37 to 45% but the initial activity was restored again in 4–24 hours. The mechanism of this adaptation is not understood but it undoubtedly is of agronomic significance in situations where the soil oxygen tension is reduced by conditions such as flooding or soil compaction.

The effects of water, both its excess and deficiency, on nitrogen fixation by soybean nodules has been extensively investigated. Sprent (1972c) working with PORTAGE soybeans has shown that water supply is probably the major environmental factor affecting nitrogen fixation. Maximum nitrogen fixation occurred with the soil near its field capacity, but was reduced at soil moisture levels both above and below this value. Dry soil surrounding the nodules is not detrimental to their activity as long as water is made available to the plant through deeper portions of the root system (Hume et al., 1976). By studying gradually drying detached soybean nodules (Sprent, 1971) it was seen that loss of more than 20% of the water present in fully turgid nodules caused nitrogen fixation to cease and nodule respiration to become very low.

Such effects of dehydration were irreversible. At less than 20% dehydration, proportionate decreases in N_2 fixation and nodule respiration occurred, but these changes were reversible upon watering. Her results suggest that there may be considerable day to day variation in the nitrogen fixed by soybean nodules due to variations in water supply.

Mild water stress causes structural changes (disruption of plasmodesmata) in the outer, noninfected, cortical cells of soybean nodules (Sprent, 1972a,b). Such damage can be brought about not only by excessive atmospheric transpirational demand, but can also be induced osmotically by hypotonic solutions of nonelectrolytes in contact with the roots and nodules.

The decreased nitrogen fixation of water-stressed nodules has been attributed to decreased nodule respiration caused by an insufficiency of oxygen in the bacteroids (Pankhurst and Sprent, 1975). It is thought that such an insufficiency of oxygen is due to the development of a physical barrier to oxygen diffusion into the nodule and/or due to an altered affinity of leghemoglobin for oxygen in water-stressed nodules. The proposal is supported by the observation that the inhibitory effects of mild water stress on nodule nitrogen fixation and respiration can be reversed by increasing the ambient oxygen concentration around the nodules. The acetylene reduction activity of sliced nodules, nodule bries, or bacteroid preparations from water-stressed and nonstressed nodules were found to be almost identical, presumably because in such preparations the availability of oxygen is less dependent upon diffusion through nodular tissues. When Pankhurst and Sprent (1976) studied the temperature responses of respiration in stressed and nonstressed soybean and French bean nodules, they found the temperature responses (Q_{10} values) to be indicative of a diffusion-limited process.

In contrast, Huang et al. (1975a,b), also working with soybean nodules, concluded that the inhibitory effect of water stress on acetylene reduction (nitrogen fixation) is due to reduced photosynthetic rates in the leaves of stressed plants. By making simultaneous measurements of acetylene reduction (N_2 fixation) in the nodules and photosynthesis in the leaves of soybean plants at various times after water was withheld from the soil, they demonstrated that acetylene reduction dropped sharply at a water potential of -12 bars and stopped at about -25 bars. The photosynthetic response to water potential followed an almost identical curve. The inhibition of acetylene reduction by low water potential could be mimicked by depriving the shoots of carbon dioxide, thus inhibiting photosynthesis; water stress-induced inhibition of acetylene reduction could be partially reversed by increasing the carbon dioxide concentration around the shoots.

Such a photosynthetic mechanism of water stress effects on nodule activity can obviously not apply to observations on the effect of desiccation of isolated

3. Assimilation

nodules (Sprent, 1971), so the relative roles of the two proposed mechanisms remain to be clarified.

Most reports on the effect of mineral nutrition on nitrogen fixation relate to effects of the availability of combined forms of nitrogen to the plant. Potassium has been reported to stimulate nitrogen fixation in nodules of *Vicia faba* (Mengel et al., 1974) by virtue of stimulating photosynthesis, but no such reports have been found for soybeans.

The presence of combined nitrogen (NO_3^-) in the environment has been shown to inhibit both nodule initiation and nodule development. When combined nitrogen (as ammonium nitrate) was provided to only one part of a soybean root system but not the other, it was found that both nodule initiation and nodule growth were inhibited in that half of the root which was exposed to fixed nitrogen (Hinson, 1975). In the other half, nodule initiation was not affected but nodule growth was decreased.

Combined nitrogen, of course, stimulates plant growth so that in spite of its reported effect of inhibiting nodule initiation, one can often demonstrate an advantage of small amounts of combined nitrogen during early seedling growth on the number of nodules formed per plant (Hatfield et al., 1974). This is presumably a root growth response to combined nitrogen by the young seedling before it develops its symbiotic nitrogen assimilation capability.

Symbiotic N_2 fixation is closely dependent upon respiratory substrates supplied by the photosynthetic process in the leaves. In peas it has been estimated that 32% of the total carbon fixed photosynthetically in the top is used by the nodules (Minchin and Pate, 1973). It can be calculated (Ching et al., 1975) that at least 12 moles of ATP are required in the nodules for every mole of N_2 fixed. Because of this, considerable environmental control of N_2 fixation is exerted indirectly by environmental effects on photosynthesis.

4. Diurnal and Seasonal Patterns

There are somewhat conflicting reports in the literature on diurnal variation in soybean nodule activity. Mague and Burris (1972) have reported a pronounced diurnal variation in acetylene reduction rates of field-grown soybeans (unspecified variety) with an activity peak at 5 P.M. The activity was correlated with both light intensity and air temperature, but not with soil temperature. Other workers (Fishbeck et al., 1973; Thibodeau and Jaworski, 1975), however, have not been able to show a significant diurnal variability in acetylene reduction by soybean nodules. The latter should not be construed to indicate that nitrogen fixation is not dependent upon photosynthesis, however, since the size of the photosynthate pools in the nodules was not reported.

Temperature effects on nodule activity were investigated by Kuo and Boersma (1971) who found an optimum soil temperature of 23.9°C over a range of soil water potentials. Pankhurst and Sprent (1976) however, found a broad optimum temperature of between 15° and 30°C for nitrogen fixation by nodules of PORTAGE soybeans, and this optimum temperature range decreased with water stress.

Air temperature may have as great, if not greater, effect than soil temperature on nodule activity (Mague and Burris, 1972), perhaps because of its effect on translocation of carbohydrates from the leaves to the nodules.

The seasonal pattern of soybean nodule activity has been investigated by several workers (Lawn and Brun, 1974; Klucas, 1974; Thibodeau and Jaworski, 1975; Weil and Ohlrogge, 1975; Harper, 1974). In all of these reports it was found that N_2 fixation per plant was slow early in the season but then increased markedly and reached a peak sometime after flowering, whereupon it dropped dramatically during the early seed-filling period. Lawn and Brun (1974) have ascribed the steep drop-off in activity to competition between the developing seeds and the nodules for the available photosynthate. By experimentally altering the relative strengths of sources and sinks (by supplemental light, shade, depodding and defoliation treatments) during the pod-filling period, they were able to modify the rates of acetylene reduction activity in the nodules as would be predicted by such a hypothesis of substrate limitation.

As discussed later, the nitrate reductase assimilation system in soybeans also shows a peak followed by a decline, but this decline occurs about 2 or 3 weeks earlier than the decline in nodule activity (Harper, 1974; Thibodeau and Jaworski, 1975). The nitrate reductase activity decline has been ascribed by Thibodeau and Jaworski (1975) to a decreased rate of nitrate absorption by the roots, occasioned by competition between roots and nodules for photosynthetic substrates.

There emerges the picture of three sequential competitive sinks for photosynthate: the nitrate absorption in the roots, the fixation of N_2 in the nodules, and the filling of the seeds in the developing pods. It would obviously be of great interest to gain a better understanding of the physiological mechanism underlying such a shifting relationship between sources and sinks in the soybean plant.

B. Nitrate Reduction

1. *The Site of Reduction*

As discussed earlier, both of the nitrogen assimilatory systems appear to be essential for maximum yield of soybeans (Harper, 1974), with the nitrate reductase system contributing its maximum slightly earlier in plant development than does the N_2 fixation system.

Soybeans can absorb nitrogen from the soil in three forms: as nitrate ions, as ammonium ions, or as organic compounds such as amino acids or urea. Of these, nitrate is by far the most important because of its relative stability in the soil and because the other forms tend to be converted to nitrate by the action of various soil microorganisms.

Once absorbed by the root system, the nitrate ion must be metabolically reduced to the level of amino nitrogen before it is of any value to the plant. The reduction of the nitrate ions, which in other species can occur both in the root and in the leaves, appears to occur almost exclusively in the leaves of the soybean plant (Beevers and Hageman, 1969).

The enzyme nitrate reductase which is the rate-limiting enzyme in the multistep reduction, is an induced enzyme, that is, its synthesis is stimulated by the presence of the substrate, the nitrate ion. The enzyme is a flavoprotein containing flavine adenine dinucleotide (FAD) and molybdenum as its prosthetic group. The reaction that it catalyzes is the reduction of nitrate to nitrite which it does at the expense of a reduced coenzyme, either NADH or NADPH. The relative role of these two coenzymes in soybean leaves is not clear (Jolly et al., 1975).

The nitrite ion produced in the reaction is then further reduced by nitrite reductase to yield certain labile intermediates which are quickly reduced to ammonia, which is further metabolized to amino acids.

The rate-limiting nitrate reductase activity in leaves is stimulated by both the intensity and duration of light. It has been suggested that this is in response to changes in cell membrane permeabilites influencing nitrate absorption (Beevers and Hageman, 1969), although light is also known to stimulate the induction of nitrate reductase activity by increased protein synthesis (Travis et al., 1970).

Soybean leaves display a clear diurnal variation in nitrate reductase activity (Harper and Hageman, 1972) which is, at least partially, related to the effect of light on enzyme induction, but which is also modified by a marked temperature sensitivity of the activity during darkness as well as by an energy limitation (Nicholas et al., 1976a,b).

Harper and Hageman (1972) have documented the changes in nitrate reductase activity occurring in the leaves of field-grown BEESON soybeans. They find that the total nitrate reductase activity per plant per hour shows a sharp peak about 60 days after emergence and then gradually declines and becomes very low at maturity. At any given time within the canopy there was also a distinct activity profile among the various leaves on the plant. In young plants, nitrate reductase activity per leaf per hour was greatest in the uppermost fully expanded leaf and then declined steadily at lower leaf positions, while in older plants, maximum activity per leaf per hour shifted to lower leaves in the canopy. Nustad (1976) has observed very similar seasonal and canopy profiles in CHIPPEWA 64 and CLAY soybeans, and has recorded a high positive correlation between nitrate reductase activity and net assimilation rate in different leaf layers of field-grown canopies. She also found that removal of the reproductive sinks on these plants stimulated nitrate reductase activity per gram of leaf tissue per hour, thus reemphasizing the interdependency of different sink activities on a common source of photosynthate.

2. Environmental Effects

Environmental effects on the nitrate reductase activity of soybean leaves have not been investigated extensively. Nicholas *et al.* (1976a) have found that following a period of darkness, nitrate reductase activity of CALLAND soybean leaves increased from the onset of illumination and was proportional to light intensity. Their data suggests that light was affecting nitrate reductase activity not only by stimulating enzyme induction, but also by providing more reductant (NADH) or by increasing the availability of nitrate ions to the leaves. The latter could be caused by greater nitrate absorption or translocation within the plant.

The effects of temperature on nitrate reduction in soybean leaves has been investigated by Nicholas *et al.* (1976a) who found that the rate of decline of *in vivo* nitrate reductase activity in the dark is markedly temperature-sensitive. Cool night temperatures (20°C) considerably delayed the normal night time decline in activity. In contrast, the recovery of activity in the light the following morning was temperature independent.

IV. SUMMARY

The emerging picture of the carbon and nitrogen assimilatory process in soybeans is one of great complexity. Not only is there a strong environmental influence on each of the systems, which causes distinct diurnal and seasonal activity profiles, but there is also a strong interaction between the two systems. This interaction is caused by the fact that nitrogen assimilation must depend on carbon assimilation for its chemical driving force as well as for carbon compounds in the production of organic nitrogenous material.

The picture is further complicated by the fact that there are two distinct nitrogen assimilating systems. The relationship between these two systems is determined largely by the availability of nitrogen in the soil, and is such that they compliment each other in fulfilling the demand for amino nitrogen by the vegetative and reproductive sinks in the plant.

3. Assimilation

These sinks, specifically the reproductive sink, are very competitive for photosynthate, and this competition may be the cause of the eventual decline in nitrogen assimilation as the pods fill.

While a great deal is known about soybean assimilation, much remains to be learned. With the present scarcity of energy resources available to agriculture, we need to take increasing advantage of the soybean plant's ability to assimilate its own nitrogen from the abundant N_2 gas in the atmosphere. This apparently must be done at the expense of the plant's carbon economy. We consequently need to know a great deal more about how the plant itself manages this relationship. If we could learn more about the mechanisms governing the interaction between the carbon and nitrogen assimilation systems, we perhaps could influence it to our advantage by manipulating the plant either genetically, culturally, or chemically.

ACKNOWLEDGMENT

This chapter is a contribution of the Department of Agronomy and Plant Genetics (Univ. of Minnesota, St. Paul, Minnesota 55108). Paper No. 9939, Scientific Journal Series, Minnesota Agricultural Experiment Station.

REFERENCES

Balegh, S. E., and Biddulph. O. (1970). *Plant Physiol.* **46**, 1–5.
Beevers, L., and Hageman, R. H. (1969). *Annu. Rev. Plant Physiol.* **20**, 495–522.
Bergersen, F. J. (1970). *Aust. J. Biol. Sci.* **23**, 1015–1025.
Bernard, R. L., and Weiss, M. G. (1973). *In* "Soybeans: Improvement, Production and Uses" (B. E. Caldwell *et al.*, eds.), pp. 117–154. Am. Soc. Agron., Madison, Wisconsin.
Bethlenfalvay, J. J., and Phillips, D. A. (1977). *Plant Physiol.* **60**, 419–421.
Beuerlein, J. E., and Pendleton, J. W. (1971). *Crop Sci.* **11**, 217–219.
Bjorkman, O. (1968). *Physiol. Plant.* **21**, 1–10.
Bowes, G. W., Ogren, W. L., and Hageman, R. H. (1972). *Crop Sci.* **12**, 77–79.
Boyer, J. S. (1965). *Plant Physiol.* **40**, 229–234.
Boyer, J. S. (1970a). *Plant Physiol.* **16**, 233–235.
Boyer, J. S. (1970b). *Plant Physiol.* **46**, 236–239.
Brun, W. A., and Cooper, R. L. (1967). *Crop Sci.* **7**, 451–454.
Burris, R. H., Eppling, F. J., Wahlin, H. B., and Wilson, P. W. (1943). *J. Biol. Chem.* **148**, 349–357.
Buttery, B. R., and Buzzell, R. I. (1977). *Can. J. Plant Sci.* **57**, 1–5.
Chartier, P., Chartier, M., and Catsky, J. (1970). *Photosynthetica* **4**, 48–57.
Ching, T. M., Hedtke, S., Russell, S. A., and Evans, J. (1975). *Plant Physiol.* **55**, 796–798.
Ciha, A. J., and Brun, W. A. (1975). *Crop Sci.* **15**, 309–313.
Clauss, H., Mortimer, D. C., and Gorham, P. R. (1964). *Plant Physiol.* **39**, 269–273.
Cook, G. M., and Evans, L. T. (1976). *In* "Transport and Transfer Processes in Plants" (I. F. Wardlaw and J. B. Passioura, eds.), pp. 393–400. Academic Press, New York.

Criswell, J. G., Havelka, U. D., Quebedeaux, B., and Hardy, R. W. F. (1976). *Plant Physiol.* **58**, 622-625.
Criswell, J. G., Havelka, U. D., Quebedeaux, B., and Hardy, R. W. F. (1977). *Crop Sci.* **17**, 39-44.
Curtis, P. E., Ogren, W. L., and Hageman, R. H. (1969). *Crop Sci.* **9**, 323-327.
Dilworth, M. J. (1966). *Biochim. Biophys. Acta* **127**, 285-294.
Dornhoff, G. M., and Shibles, R. M. (1970). *Crop Sci.* **10**, 42-45.
Doss, B. D., Pearson, R. W., and Rogers, H. T. (1974). *Agron. J.* **66**, 297-299.
Dreger, R. H., Brun, W. A., and Cooper, R. L. (1969). *Crop Sci.* **9**, 429-431.
Duncan, W. G. (1971). *Crop Sci.* **11**, 482-485.
Egli, D. B., Pendelton, J. W., and Peters, D. B. (1970). *Agron. J.* **62**, 411-414.
Evans, L. T. (1976). *In* "Transport and Transfer Processes in Plants" (J. F. Wardlaw and J. B. Passioura, eds.), pp. 1-13. Academic Press, New York.
Fishbeck, K., Evans, H. J., and Boersma, L. L. (1973). *Agron. J.* **65**, 429-433.
Flinn, A. M. (1974). *Physiol. Plant.* **31**, 275-278.
Forrester, M. L., Krotkov, G., and Nelson, C. D. (1966). *Plant Physiol.* **41**, 422-427.
Gallaher, R. N., Ashley, D. A., and Brown, R. H. (1975). *Crop Sci.* **15**, 55-59.
Gastra, P. (1959). *Meded. Landbouwhogesch. Wageningen* **59**, 1-68.
Ghorashy, S. R., Pendleton, J. W., Peters, D. B., Boyer, J. S., and Beuerlein, J. E. (1971). *Agron. J.* **63**, 674-676.
Gibson, A. H. (1965). *Aust. J. Biol. Sci.* **19**, 499-515.
Gibson, A. H. (1974). *R. Soc. N.Z., Bull.* **12**, 13-22.
Hansen, P. (1971). *Physiol. Plant.* **25**, 181-183.
Hanway, J. J., and Weber, C. R. (1971). *Agron. J.* **63**, 406-408.
Hardman, L. L., and Brun, W. A. (1971). *Crop Sci.* **11**, 886-888.
Hardy, R. F. W., and Havelka, U. D. (1976). *In* "Symbiotic Nitrogen Fixation in Plants" (P. S. Nutman, ed.), pp. 356-362. Cambridge Univ. Press, London and New York.
Hardy, R. F. W., Holsten, R. D., Jackson, E. K., and Burns, R. C. (1968). *Plant Physiol.* **43**, 1185-1207.
Harper, J. E. (1974). *Crop Sci.* **14**, 255-260.
Harper, J. E., and Hageman, R. H. (1972). *Plant Physiol.* **49**, 146-154.
Harper, L. A., Baker, D. N., Box, J. E., Jr., and Hesketh, J. D. (1973). *Agron. J.* **65**, 7-11.
Hatfield, J. L., Egli, D. B., Leggett, J. E., and Peaslee, D. E. (1974). *Agron. J.* **66**, 112-115.
Hicks, D. R., and Pendleton, J. W. (1969). *Crop Sci.* **9**, 435-437.
Hinson, K. (1975). *Agron. J.* **67**, 799-804.
Hodgkinson, K. C. (1974). *Aust. J. Plant Physiol.* **1**, 561-578.
Huang, C. Y., Boyer, J. S., and Vanderhoef, L. N. (1975a). *Plant Physiol.* **56**, 222-227.
Huang, C. Y., Boyer, J. S., and Vanderhoef, L. N. (1975b). *Plant Physiol.* **56**, 228-232.
Hume, D. J., and Criswell, J. G. (1973). *Crop Sci.* **13**, 519-524.
Hume, D. J., Criswell, J. G., and Stevenson, K. R. (1976). *Can. J. Plant Sci.* **56**, 811-815.
Jarvis, P. G. (1971). *In* "Plant Photosynthetic Production, Manual of Methods" (Z. Sestak, J. Catsky, and P. G. Jarvis, eds.), pp. 566-631. Junk Publ., The Hague.
Jeffers, D. L., and Shibles, R. M. (1969). *Crop Sci.* **9**, 762-764.
Johnston, T. J., Pendleton, J. W., Peters, D. B., and Hicks, D. R. (1969). *Crop Sci.* **9**, 577-581.
Jolly, S. O., Campbell, W. H., and Tolbert, N. E. (1975). *Plant Physiol.* **56**, Suppl. 73.
Jones, H. G., and Slatyer, R. O. (1972). *Plant Physiol.* **50**, 283-288.
Keck, R. W., Dilley, R. A., and Ke, B. (1970). *Plant Physiol.* **46**, 699-704.
Klucas, R. V. (1974). *Plant Physiol.* **54**, 612-616.
Koller, H. R., and Dilley, R. A. (1974). *Crop Sci.* **14**, 779-782.
Kriedmann, P. E., Loveys, B. R., and Downton, W. J. S. (1975). *Aust. J. Plant Physiol.* **2**, 253-267.

Kuo, T., and Boersma, L. (1971). *Agron. J.* **63**, 901–904.
Laing, W. A., Ogren, W. L., and Hageman, R. H. (1974). *Plant Physiol.* **54**, 678–685.
Lawn, R. J., and Brun, W. A. (1974). *Crop Sci.* **14**, 11–16.
Lenz, R. (1974). *Proc. Int. Hortic. Congr., 19th, 1974* pp. 155–166.
Loveys, B. R., and Kriedeman, P. E. (1974). *R. Soc. N.Z., Bull.* **12**, 781–787.
Mague, T. H., and Burris, R. H. (1972). *New Phytol.* **71**, 275–286.
Mederski, H. J., Chen, L. H., and Curry, R. B. (1975). *Plant Physiol.* **55**, 589–593.
Mengel, K., Haghparast, M., and Koch, K. (1974). *Plant Physiol.* **54**, 535–538.
Minchin, F. R., and Pate, J. S. (1973). *J. Exp. Bot.* **24**, 259–271.
Misken, K. E., Rasmusson, D. C., and Moss, D. N. (1972). *Crop Sci.* **12**, 780–783.
Mondal, M. H., Brun, W. A., and Brenner, M. L. (1978). *Plant Physiol.* **61**, 394–397.
Nafziger, E. D., and Koller, H. R. (1976). *Plant Physiol.* 57, 560–563.
Neals, T. F., and Incoll, L. D. (1968). *Bot. Rev.* **34**, 107–125.
Nicholas, J. C., Harper, J. E., and Hageman, R. H. (1967a). *Plant Physiol.* **58**, 731–735.
Nicholas, J. C., Harper, J. E., and Hageman, R. H. (1967b). *Plant Physiol.* **58**, 736–739.
Nustad, L. L. (1976). Ph.D. Thesis, University of Minnesota, St. Paul.
Ogren, W. L., and Bowes, G. (1971). *Nature (London), New Biol.* **230**, 159–160.
Pallas, J. E., Jr. (1973). *Crop Sci.* **13**, 82–84.
Pankhurst, C. E., and Sprent, J. I. (1975). *J. Exp. Bot.* **26**, 287–304.
Pankhurst, C. E., and Sprent, J. I. (1976). *J. Exp. Bot.* **27**, 1–9.
Pate, J. S. (1976). *In* "Transport and Transfer Processes in Plants" (J. F. Wardlaw and J. B. Passioura, eds.), pp. 447–462. Academic Press, New York.
Pirl, D. L. Jr. (1976). M.S. Thesis, University of Minnesota, St. Paul.
Plaut, Z. (1971). *Plant Physiol.* **48**, 591–595.
Quebedeaux, B., and Chollet, R. (1975). *Plant Physiol.* **55**, 745–748.
Sakamoto, C. M., and Shaw, R. H. (1967a). *Agron. J.* **59**, 7–9.
Sakamoto, C. M., and Shaw, R. H. (1967b). *Agron. J.* **59**, 73–75.
Sanders, T. H., Ashley, D. A., and Brown, R. H. (1977). *Crop Sci.* **17**, 548–550.
Schollhorn, R., and Burris, R. H. (1966). *Fed. Proc., Fed. Am. Soc. Exp. Biol.* **25**, 70 (abstr.).
Schubert, K. R., and Evans, H. J. (1976). *Proc. Natl. Acad. Sci. U.S.A.* **73**, 1207–1211.
Shibles, R. M., and Weber, C. R. (1965). *Crop Sci.* **5**, 575–577.
Shibles, R. M., and Weber, C. R. (1966). *Crop Sci.* **6**, 55–59.
Sinclair, T. R., and de Wit, C. T. (1975). *Science* **189**, 565–567.
Sinclair, T. R., and de Wit, C. T. (1976). *Agron. J.* **68**, 319–324.
Sprent, J. I. (1971). *New Phytol.* **70**, 9–17.
Sprent, J. I. (1972a). *New Phytol.* **71**, 443–450.
Sprent, J. I. (1972b). *New Phytol.* **71**, 451–460.
Sprent, J. I. (1972c). *New Phytol.* **71**, 603–611.
Starnes, W. J., and Dadley, H. H. (1965). *Crop Sci.* **5**, 9–11.
Tamas, I. A., Schwartz, J. W., Hagin, J. M., and Simmonds, R. (1974). *R. Soc. N.Z., Bull.* **12**, 261–268.
Thaine, R., Ovenden, S. L., and Turner, J. S. (1959). *Aust. J. Biol. Sci.* **12**, 349–372.
Thibodeau, P. S., and Jaworski, E. G. (1975). *Planta* **127**, 133–147.
Thorne, J. H., and Koller, H. R. (1974). *Plant Physiol.* **54**, 201–207.
Thrower, S. L. (1962). *Aust. J. Biol. Sci.* **15**, 629–649.
Travis, R. L., Huffaker, R. C., and Key, J. L. (1970). *Plant Physiol.* **46**, 800–805.
Treharne, K. J., Stoddard, J. L., Pughe, J., Paranhothy, K., and Wareing, P. F. (1970). *Nature (London)* **228**, 129–131.
Turner, W. B., and Bidwell, R. G. S. (1965). *Plant Physiol.* **40**, 446–451.
Upmeyer, D. J., and Koller, H. R. (1973). *Plant Physiol.* **51**, 871–874.
Wareing, P. F., Khalifa, M. M., and Treharne, K. J. (1968). *Nature (London)* **220**, 453–459.

Weber, C. R. (1966a). *Agron. J.* **58**. 43–46.
Weber, C. R. (1966b). *Agron. J.* **58,** 46–49.
Weil, R. R., and Ohlrogge, A. J. (1975). *Agron. J.* **67,** 487–490.
Wickliff, J. L., and Aronoff, S. (1962). *Plant Physiol.* **37,** 590–594.
Woolley, J. T. (1964). *Agron. J.* **56,** 569–571.

4

Agronomic Characteristics and Environmental Stress

D. KEITH WHIGHAM and HARRY C. MINOR

I.	Introduction	78
II.	Light	78
	A. Response to Day Length	81
	B. Response to Light Intensity	87
	C. Response to Light Quality	88
	D. Interaction between Light and Temperature	88
III.	Temperature	89
	A. Effect of Temperature on Growth and Development	89
	B. Effect of Temperature on Nitrogen Fixation	91
	C. Effect of Temperature on Physiological Processes	92
	D. Effect of Temperature on Seed Quality	93
	E. Effect of Temperature on Protein and Oil	94
	F. Effect of Temperature on Pests	95
IV.	Water	97
	A. Effect of Mulching	98
	B. Effect of Water Deficit	99
	C. Effect of Excess Water	100
	D. Effect of Atmospheric Humidity	102
V.	Wind	102
	A. Effect of Wind	102
	B. Effect of Windbreaks	104
VI.	Pests	105
	A. Birds and Rodents	105
	B. Diseases	106
	C. Insects	109
	D. Weeds	113
VII.	Conclusions	115
	References	116

I. INTRODUCTION

The agronomic characteristics of soybeans are those plant characters which are used to describe a soybean plant and its development. Examples include grain yield, days to flower, plant height, leaf area, oil and protein content, seed quality, and many others. These characteristics are modified in their expression by environmental conditions, which vary between seasons, locations, and years. Any major variation in the environment may result in a stress on the plant. Shading, extremely high temperatures, inadequate nutrients, high intensity wind, and damage by pests are types of stress.

This chapter will attempt to describe some of the effects that different environmental variables have on soybean plant characteristics. How the soybean plant responds to environmental variables ultimately determines the areas of the world where soybeans can be produced successfully.

II. LIGHT

The early introductions of soybeans into the United States were recognized to differ greatly in their growth behavior and adaptation to different

TABLE I

Relationship of Latitude to Effective Photoperiod, by Calendar Date, at Various Latitudes in the Northern Hemisphere[a]

| Degrees latitude | Effective photoperiod[b] ||||||||
| | Dec. 20 || Sept. 20–March 20 || June 20 || Maximum difference ||
	Hours	Minutes	Hours	Minutes	Hours	Minutes	Hours	Minutes
10°	12	20	12	52	13	30	1	10
15	12	00	12	55	13	50	1	50
20	11	42	12	58	14	10	2	28
25	11	28	13	00	14	30	3	02
30	11	10	13	00	15	00	3	50
35	10	52	13	08	15	40	4	48
40	10	30	13	12	16	15	5	45
45	10	08	13	12	17	03	6	55

[a] Modified from Hartwig, 1970.
[b] Duration of daylight from beginning of civil twilight in the morning to the end of civil twilight in the evening.

4. Agronomic Characteristics and Environmental Stress

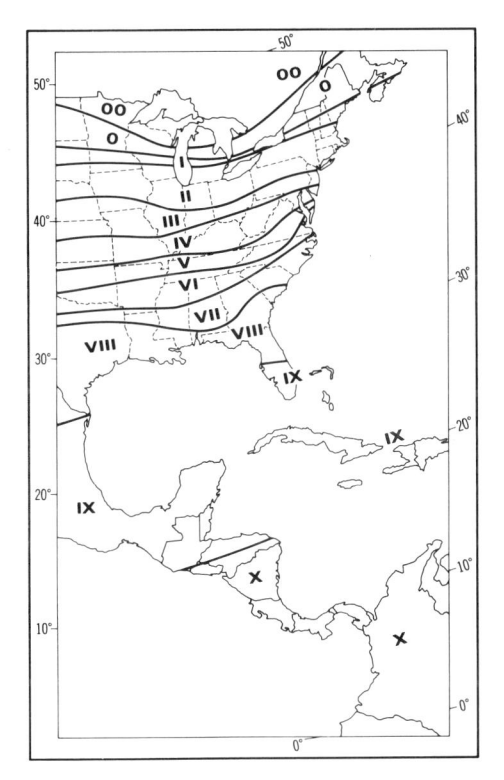

Fig. 1. Zones of best adaptation for cultivars of soybean Maturity Groups 00 through X.

environments. Many of the observed differences were later associated with responsiveness to photoperiod, which affects the time to flowering and maturity. These differences were utilized by plant breeders to develop cultivars adapted to rather narrow latitudinal belts in the United States. The belts are loosely defined environmental limits primarily determined by day length (Fig. 1). The maximum range in day length at 10° latitude is 1 hour 10 minutes but at 45° latitude is 6 hours 55 minutes (Table I). During the growing season only, the range in day length at 45° latitude is approximately 3 hours. For this reason, the belts are rather wide at low latitudes and are quite narrow at high latitudes.

Light strongly influences the morphology of the soybean plant by causing changes in the time of flowering and maturity, which result in differences in plant height, pod height, leaf area, lodging, and numerous other plant characteristics including grain yield. Light is also essential to drive the photosynthetic mechanism which influences N fixation, total dry matter production, grain yield, and the many other characteristics dependent on the production of photosynthate. The interaction between the photomorphogenic effects and those due to photosynthetic output is complex. For example, plants which flower early, due to short day length, do not usually develop normal plant height or leaf area. Maturation will also be hastened and a lower grain yield than normal may be produced because of the reduced photosynthetic input.

Based on the recognition of adaptation belts, cultivars have been classified in maturity groups depending on their responsiveness to the environment. Cultivars are usually placed in one of twelve groups from Maturity Group 00 to X. Table II identifies some cultivars in the various maturity groups. Maturity Group 00 is the designation for the earliest cultivars; these are adapted to environments normally associated with latitudes of 50° or greater. The latest maturity group is X which includes cultivars developed for production in tropical environments at low latitudes. Cultivars classified in the same maturity group often have different development rates and may mature at somewhat different times. The range in the number of days to maturity within the same group may, over a series of environments, average as much as three weeks. The system of maturity group classification permits approximate identification of the zone of best adaptation of cultivars. However, use of the system outside of North America may be misleading unless the environment is similar to that in which the cultivar was developed and classified.

TABLE II

Some Cultivars Classified by Maturity Group in North America

Maturity group	Cultivar name	Maturity group	Cultivar name
00	ALTONA	V	FORREST
	PORTAGE		HILL
0	MERIT	VI	DAVIS
	SWIFT		LEE 74
I	HARK	VII	BRAGG
	HODGSON		SEMMES
II	BEESON	VIII	HARDEE
	CORSOY		IMPROVED PELICAN
III	CALLAND	IX	JUPITER
	WILLIAMS		
IV	CLARK 63	X	(Primarily experimental lines)
	KENT		

A. Response to Day Length

Soybeans have been recognized as quantitative short-day plants (Garner and Allard, 1920, 1923). The length of the dark period is the controlling factor in eliciting photoperiodic responses. Other environmental factors may influence photoperiodic responses but do not have as great an effect as the number of hours of darkness.

Cultivars differ in their response to day length (Fig. 2). Differences in day length result in responses in terms of the number of days to flower, the number of days to maturity, plant height, seed weight, pod number, branch number, node number, and others. Flowering occurs when the day length becomes shorter than the critical value for the cultivar. Plants 1 to 2 weeks old do not respond to photoperiodic treatment, but older plants receiving the same treatment respond by the initiation of flower buds.

Late maturing cultivars are more sensitive to photoperiod than are early maturing cultivars. Studies with the earliest maturing lines—those adapted to the northernmost production areas in the United States—indicate that the number of days to flower of most of these strains is unaffected by day length. The number of days to flower for most cultivars of later maturity groups is increased by longer photoperiods. In the United States flowering is delayed when cultivars normally grown in Arkansas and Mississippi are planted in Iowa or Minnesota. In fact, if a Maturity Group IX cultivar is planted at 50° latitude, frost may kill the plants before they flower. The opposite is true

Fig. 2. Relative plant height and maturity of soybean cultivars WILLIAMS (left) and JUPITER (right) bordered by local cultivars in Indonesia.

Fig. 3. Early maturing cultivar (left) and a late maturing cultivar (right) in Brazil.

when cultivars from the north are planted south of their adapted environment. A Maturity Group I cultivar planted in Louisiana at 30° latitude will flower and mature much earlier than when grown in Iowa. Yield may also be adversely affected. Figure 3 compares development of a cultivar of temperate origin with that of one adapted to low latitudes, when both are grown near the equator.

Tropical cultivars are significantly later in flowering during 12- and 14-hour photoperiods than temperate cultivars (Byth, 1968). Increasing day length from 10 to 14 hours greatly extends the number of days to flower for the tropical cultivars, but a 16-hour photoperiod has little additional effect. The temperate cultivars exhibit a linear increase in the number of days to flower when day length is extended from 12 to 16 hours. Longer day lengths delay flowering and usually result in the development of a larger number of flowers, but the percentage of flower and pod shedding also increases. Short days following the initiation of flowering result in fewer days from flowering to maturity. There is usually a positive association between day length and plant height. When day length is changed from 8 to 16 hours, plant height, node number, internode length, leaf area, and days to flower all increase with both tropical and temperate cultivars when measured at flowering. The increased growth is due to more days before flowering rather than to increased growth rates. The number of days to maturity increases with longer day length, but part of this is due to changes in the number of days to flower. The effect of shortening the day length after flowering is manifest in fewer days to maturity.

4. Agronomic Characteristics and Environmental Stress

1. Response to Different Latitudes

The relationships between day length, latitude in the northern hemisphere, and calendar date are shown in Table I. The relationships between day length and latitude are similar for the southern hemisphere, but the calendar dates are, of course, different.

In a test of widely adapted soybean cultivars in Puerto Rico (18° N latitude), most cultivars flowered in 30 days or less after emergence (Hartwig, 1970). Table III shows the effect of latitude and day length on flowering and maturity of six United States cultivars. The cultivars were developed for production between the latitudes of 30° and 40° in the United States. When moved to the tropics, the development periods are similar for all cultivars. Flowering occurs early in response to the short day length, and the number of days from flowering to maturity is less variable among cultivars than in areas with longer day lengths. Cultivars HILL and HARDEE differ in the number of days to flower by two days at 1°, 8°, and 19° latitude, but the difference in the number of days from flowering to maturity is 6, 10, and 63

TABLE III

The Effect of Latitude and Day Length on Flowering and Maturity of United States Soybean Cultivars[a]

		\multicolumn{8}{c}{Site[b]}							
		Ecuador (1°S) (12.6)[c]		Sri Lanka (8° N) (12.9)[c]		Dominican Republic (19° N) (13.2)[c]		Israel (32° N) (13.7)[c]	
Cultivar	Maturity group	E-F[d]	F-M[e]	E-F[d]	F-M[e]	E-F[d]	F-M[e]	E-F[d]	F-M[e]
WILLIAMS	III	26	62	21	60	34	56	37	83
CLARK 63	IV	26	63	22	59	34	56	38	85
HILL	V	31	57	27	61	39	58	77	67
DAVIS	VI	30	62	28	63	38	78	90	62
BRAGG	VII	27	66	25	63	33	83	77	87
HARDEE	VIII	29	63	29	71	41	121	108	71

[a] From Whigham, 1976b.
[b] Altitude range among sites was from 44 to 200 m.
[c] Approximate day length (hours) at the time of planting (all sites planted within 15 days of April 9).
[d] Number of days from emergence (E) to flowering (F).
[e] Number of days from flowering (F) to maturity (M).

days, respectively. All of the cultivars are sensitive to day length but have a critical photoperiod longer than 13 hours.

Maximum day length differences occur on June 21 in the Northern Hemisphere. On that date, day length is almost 1.3 hours longer at 50° latitude than at 40°. A full season cultivar adapted to 40° would normally flower about July 1, when day length is 15 hours. This same photoperiod would not occur until about August 10 at 50° latitude, and flowering would be so late that the cultivar would not mature before frost.

There are situations in which cultivars produce high yields in an environment in which they are not truly adapted. At sea level and 9° N latitude in Sri Lanka, the cultivar HARDEE produced a yield of 6054 kg/ha in 1975 experiments. Many tropical locations have produced yields which exceed 4000 kg/ha (Whigham, 1976a). In most cases, the cultivars involved were not developed for production at low latitudes and many plant characteristics are different from what is considered desirable in the United States. The number of days to flowering and days to maturity are less than normal, and plant height is often only half that achieved in the areas where the cultivars were developed. The reduction in plant height, due primarily to early flowering, increases the difficulty with which the cultivars are harvested. The lowest pods are formed close to the soil surface and generally contain seeds of lower quality. Hand harvesting is required to reduce harvest losses to an acceptable level.

2. *Response to Different Planting Dates*

Seasonal changes also affect day length; therefore, at a given location, the planting date determines whether the crop will be developing during short or long days. Changes in day length are only slight at the equator and the cropping seasons are usually determined by the rain patterns. In temperate regions, temperature and rain patterns both determine the cropping season.

In Puerto Rico, soybean yields are highest when plantings are made in May and June. The lowest yields are obtained from the December and January plantings. Smaller plants are produced and the number of days to maturity is less when the crop is planted in December and January. Seed size, protein content, and oil content are not affected by the planting date. With proper timing, three crops of soybeans can be grown in Puerto Rico and other subtropical and tropical countries during one calendar year. Irrigation water may be necessary during seasons when rainfall is insufficient.

In Sri Lanka, the maximum change in day length is less than 1 hour throughout the year. Fifteen cultivars were tested at five sites in two different seasons (Whigham, 1976b). The relationship between planting dates and several agronomic characteristics of the cultivars HARDEE, HILL, and WILLIAMS are shown in Table IV. Slightly higher yields were produced by

TABLE IV

Relationship between Planting Dates[a] and Several Agronomic Characteristics of Three Cultivars at Five Locations[b] in Sri Lanka[c]

	Cultivars					
	HARDEE		HILL		WILLIAMS	
Characteristics	Apr–May	Oct–Nov	Apr–May	Oct–Nov	Apr–May	Oct–Nov
Yield (kg/ha)	1839	1939	1506	1638	1788	1775
Days to flower	30	27	28	26	25	24
Days from flowering to maturity	65	61	53	51	63	57
Plant height (cm)	32	32	30	35	49	39
Pods per plant	44	21	27	18	23	14
Seed weight (g/100)	17.3	17.8	16.5	16.4	20.8	20.5

[a] Difference in day length between planting dates was approximately ½ hour.
[b] Locations were Alutharama, Angunukolapalessa, Gannoruwa, Maha Illuppallama, and Ratmalagara.
[c] From Whigham, 1976b.

HARDEE and HILL during the October–November season, but WILLIAMS showed essentially no difference in yield between the seasons. The difference in days to flower was small for each cultivar, but more days were required before flowering in the season with the longer day lengths. The number of days from flowering to maturity varied among seasons. Differences in cultivar response to each season are apparent. HILL averaged 52 days from flowering to maturity compared to 63 and 60 days for HARDEE and WILLIAMS, respectively. WILLIAMS, an indeterminate type, was 10 cm taller during the season with the longer day lengths. The number of pods per plant was considerably higher for each cultivar during the longer day length season. Seed weight was not greatly affected by season, but seed weight was positively associated with yield of each cultivar. The day length of approximately 13 hours is less than the critical day length of the three cultivars discussed. Differences due to planting date are attributed to environmental variables other than day length.

Because soybean cultivars do not all have the same critical day length, the effect of planting date on the number of days to flowering and days to maturity will be different for different cultivars. Figure 4 illustrates the variation among cultivars when planted at approximately 6-week intervals in Puerto Rico (Minor, 1976). CLARK 63 and WILLIAMS were insensitive to the changes in day length which resulted from different planting dates. JUPITER,

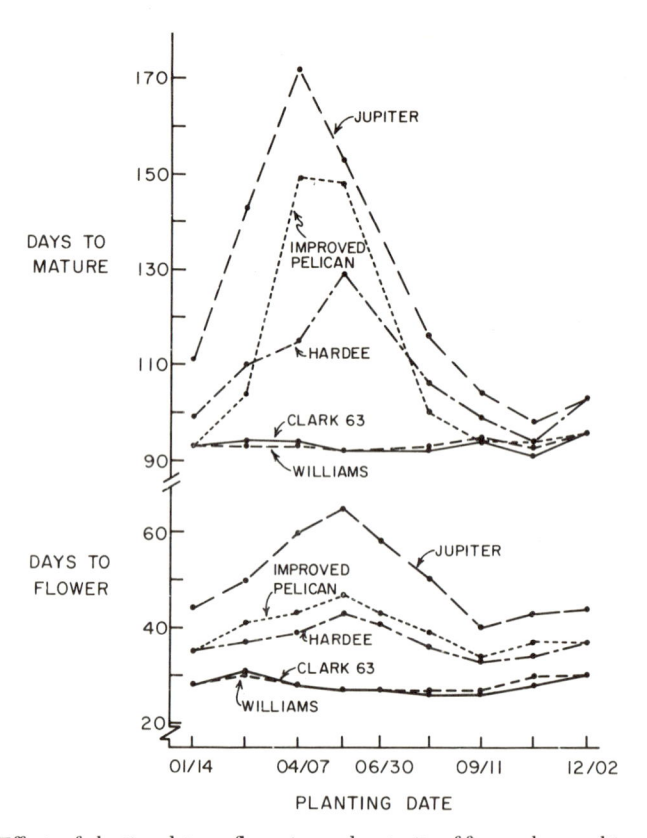

Fig. 4. Effects of planting date on flowering and maturity of five soybean cultivars, Isabela, Puerto Rico, 18° N latitude (Minor, 1976).

IMPROVED PELICAN, and HARDEE were greatly affected by the differences in day length, especially when planted so that flowering would occur near the longest day of the year. The highest yields, however, were not produced during the longest growing season for any of the late maturing cultivars. These results indicate that other environmental factors interact with day length to determine the yield potential of each cultivar.

In California, Abel (1961) found that the earliest cultivar flowered in approximately the same number of days in all biweekly plantings between May 2 and August 2. Late cultivars flowered in considerably fewer days in successive planting dates. The period from flowering to maturity for the late cultivars was unaffected by planting date, but the period was reduced for the earliest cultivars. The total number of days from planting to maturity was reduced in all cultivars. Highest yields were produced from the May plantings and the intermediate and late-maturing cultivars. Plant height was greatest from the early May plantings and lodging decreased as planting was delayed. Seed quality increased with later plantings and seed weight decreased in plantings made after July 1.

In temperate zones, as in other regions, cultivars classified as "early" usually exhibit less variation in number of days to flower and maturity than do those classified as "late." Because the length of the growth cycle becomes progressively shorter as planting is delayed, there is less variation in date of maturity than in date of planting.

In the northern United States, maximum yield is usually obtained from full-season cultivars planted early in May. In the southern United States, late May or early June are the preferred periods for planting soybeans.

B. Response to Light Intensity

Light intensity is altered by cloud cover, haze, twilight, altitude, angle of incidence, competitive shading, and other environmental conditions. Floral initiation of BILOXI soybeans occurs if the light intensity is above 1076 lux during two consecutive 8-hour photoperiods. Intensities of less than 1076 lux do not initiate flowering. Twilight should be considered as part of the natural photoperiod. In studies with BILOXI soybeans, sensitivity to more than 1 lux during the period of twilight (2–200 lux) was found following darkness. The number of days to flower was less than normal if the soybeans were kept in darkness until the light intensity was 50 lux. Flowering was delayed more as the light intensity increased.

The level of light saturation for photosynthesis in soybean leaves depends on the light intensity of the environment in which plants are grown. In greenhouse conditions, the saturation level is reached at 20 klux. In growth chambers, the rate of photosynthesis in soybean seedlings increases with light intensities up to 43 klux. Leaves of field-grown soybeans are not saturated at 150 klux. Differences in light saturation also exist among cultivars. A controversy exists among scientists concerning the light intensity at which soybeans become light saturated. The top leaves of the soybean canopy have a higher light saturation intensity and a higher rate of photosynthesis than those lower in the canopy.

A high rate of pod abscission occurs in light intensities of approximately 5 klux. Soybean yields of 20 cultivars were reduced significantly when the light intensity was decreased by 40% during shade studies in the Philippines. The average yield reduction due to shading was 32%. The number of pods was reduced by 28% and seed weight was also decreased under shade. Differences were found between cultivars in their ability to tolerate reduced light intensity.

Reflective plastic and supplemental light were used by Johnston et al. (1969) to provide a light-rich environment for soybean plants. Compared to plants without the reflective plastic or supplemental light, light-rich plants in 50-cm rows had three times as much light as normal and those in 100-cm rows had more than five times as much light as normal. Those plants receiving additional light had more nodes, branches, pods, seeds, pods per node, seeds per node, and a higher oil content than the unenhanced plants, but seed size and protein content were decreased.

Periods of high radiation are associated with sun elevation near the zenith. Environments receiving a greater proportion of radiation from directly overhead may receive more light and produce greater yields. However, should this period of maximum potential light coincide with the rainy season, light reaching the crop may actually be reduced. When water is available for irrigation, highest yields may be produced in the dry season. During the dry season, sunlight is abundant, water can be controlled, management operations can be performed in a timely manner, and weed control is less difficult.

C. Response to Light Quality

The spectral composition of light changes as it penetrates the canopy. Green and infra-red wavelengths penetrate the soybean canopy more than red or blue wavelengths. Flowering may be prevented in soybeans if the dark period is interrupted with light of sufficient energy. The radiation which most effectively delays flowering is red light at about 6400 Å. Flowering may also be delayed by exposure to blue light at about 4800 Å, but 60 times as much energy is required to equal the inhibiting effect of red light. Far-red wavelengths of about 7350 Å will reverse the effect of red light and permit normal flowering.

Light quality may be altered when the soybeans are grown in the shade of taller plants or when grown in artificial environments. Under normal field conditions, light quality will have little if any effect on soybean characteristics.

D. Interaction between Light and Temperature

The responses of soybeans to day length are modified by temperature. Cultivars differ greatly in their photoperiod sensitivity, and temperature significantly influences those cultivars least sensitive to photoperiod. Thus, early maturing cultivars respond more to changes in temperature than to day length and late maturing cultivars respond more to changes in day length than to temperature. In the temperate regions, cool temperatures and

4. Agronomic Characteristics and Environmental Stress

longer day length are additive in delaying flowering, however, short days are more important than cool temperatures for altering the number of days to maturity. Temperatures less than 25°C tend to delay flowering regardless of day length. The shedding of flowers and pods is increased by long photoperiods and temperatures of 32°C or higher. In high temperature and long day length environments, the rate of growth and final plant height are greater than in other environments. Cultivars differ in their response to temperature as it interacts with day length. In simulated tropical environments, high night temperatures promote early vegetative growth and induce early flowering. High night temperatures may offset the delay in flowering caused by long photoperiods. In Illinois, high night temperature (29.4°C) was associated with reduced yields of soybeans. Early maturity is also associated with high night temperature.

In the central United States, the highest yields are usually associated with warmer than normal mean temperatures in June but cooler than normal temperatures in July and August. In Illinois, precipitation and maximum daily temperature from June 25 to September 20 explained 68% of the variation in soybean yields. In July and August, the maximum daytime temperatures are often too high for optimum yields.

III. TEMPERATURE

A. Effect of Temperature on Growth and Development

Temperature has an effect on many growth processes of the soybean plant. Variables such as cloud cover, albedo, angle of light incidence, altitude, wind, moisture level, and season all affect the temperature of plants. The optimum temperature for rapid germination of soybean seed is about 30°C, but this temperature is rarely reached in temperate-zone soils during stand establishment. The minimum and maximum temperatures for seed germination are about 5° and 40°C, respectively. Survival rate, dry matter accumulation, and plant height of seedlings decrease when low-moisture (6%) seeds are imbibed at 5°C. High moisture (16%) seeds behave normally under the same conditions. Both high and low moisture seeds develop normally when imbibed at 25°C. A moisture level of 13-14% is the lowest which gives adequate protection from cold (5°C) temperatures during imbibition (Hobbs and Obendorf, 1972). The loss of leachates during imbibition is a function of cold temperature. Anaerobically soaking low moisture (5%) soybean seed at 5°C greatly reduced their survival and seedling vigor. Cultivar differences are observed in tolerance to cold temperature imbibition at low moisture levels.

Seedling emergence depends on depth of planting, cultivar, and soil temperatures. Insufficient hypocotyl elongation in some cultivars results in the inability to emerge from a 10-cm depth at 25°C. Differences are also observed in the germination of cultivars over a range of temperatures from 15° to 30°C. Soybean cultivars were exposed to temperature regimes of 20°, 25°, and 30°C in laboratory studies by Gilman et al. (1973). Increased length of exposure to 25°C reduced hypocotyl elongation in seedlings of some cultivars. When grown at constant temperatures, the differences among cultivars were most evident 10 days after planting. Maximum differences in hypocotyl elongation occurred at 25°C, but some inhibition in sensitive cultivars occurred at all temperatures between 21° and 28°C. The length of exposure to temperatures between 21° and 28°C affected the extent of inhibition of hypocotyl elongation. Both temperature and genotype determined the rate of hypocotyl elongation. The optimum temperature for rapid hypocotyl elongation was about 30°C. Temperatures above 40°C do not permit seeds to germinate and 10°C reduces the rate of hypocotyl elongation considerably. Therefore, soybeans should desirably be planted when the soil temperatures are approximately 30°C for rapid seedling emergence. These temperatures are possible at the time of planting in most tropical areas, but temperate-zone soils seldom reach 30°C before midseason. The seed is often planted in soils which are approximately 20°C; this results in slow emergence. Plantings made in May during a normal year in Illinois require 5 to 7 days to emerge. Cooler than normal temperatures or less than adequate moisture may delay emergence for several weeks. Under hot, tropical conditions emergence can occur in fewer than 5 days.

In greenhouse studies, the number of degree-days per trifoliolate leaf on the main stem of soybeans was about the same over the temperature range of 12°–30°C. At 30°C the rate of trifoliolate development was greater in the early spring than in the late fall season. The supply of photosynthate may have been limiting in the late fall. Temperatures of 18°C or less did not permit pod set. Seed size was greatest when plants were grown at 27°C and the number of pods per plant was highest at 30°C. Adverse ambient temperature may prevent the production, translocation, or utilization of the flowering stimulus. Temperatures below 24°C will normally delay flowering by 2 or 3 days for each decrement of 0.5°C. Floral induction is found to be greatly inhibited at 10°C or below. Flower initiation is accelerated in the central United States when mean temperatures increase from 15° to 32°C. Temperatures above 40°C have an adverse effect on the rate of node formation, internode growth rate, and flower initiation. Heat stress of 40°–46°C results in pod abscission. Even though the growth rate declines at high temperatures, soybeans tolerate high temperatures better than corn.

Brown (1960) observed that the rate of development of soybeans was zero at 10°C and maximized at 30°C, above which the rate declined. Top growth was observed when the soil temperature was as low as 2°C; an increase in soil temperature to 7°C greatly increased the top growth rate. Soybeans are more tolerant of light frost than most annual food crops.

Several soybean cultivars were evaluated at two different altitudes (1394 and 1636 m) in Sikkim (Basnet et al., 1974). Temperature was not measured at the higher altitude but was assumed to be less than at the lower altitude site, which had an average monthly maximum of 25°C during the growing season. Yield of most cultivars was less at the high altitude site. Growth and development were retarded at the higher altitude. The interval between planting and first flowering was as much as 24 days longer at the higher altitude. Flowering of the later-maturing cultivars was delayed more than that of the early-maturing cultivars. At the higher altitude, plants were shorter, lodged less, and had fewer nodes. Seeds produced at the higher altitude were of better quality.

A study of 10 soybean cultivars at 25 tropical sites indicated that the number of days from emergence to flowering increases at higher altitudes (Whigham et al., 1978). Minimum and maximum temperatures were negatively correlated with altitude. Altitude was positively associated with plant height. Prine et al. (1964) recorded higher temperatures in seeds and pods exposed to the sun on the south side of east-west rows than in seeds elsewhere in the canopy. Greater shattering and lower moisture content in the seeds were associated with high seed temperatures.

Mulching may lower soil temperature. Contradictory results have been reported for the effect of mulching on soybean yields. In Nigeria, a rice straw mulch reduced soil temperatures and improved stand establishment about 66%. In Ghana, mulched plots of soybeans were consistently 3.5°C cooler at a depth of 2.5 cm than those not mulched during the warm period of the day. Stand establishment from poor quality seed was improved due to mulching. Mulching did not affect flowering dates nor the time of maturity. Grain yields were not increased by mulching. The planting of high quality seed reduces the chance of poor stand establishment and yield in high temperature environments.

B. Effect of Temperature on Nitrogen Fixation

When not covered by a mulch or shaded by the crop canopy, tropical soils are often high in temperature. Nodulation and nitrogen fixation in soybeans are greatly affected by soil temperature. *Rhizobium japonicum* growth is limited by temperatures in excess of 33°C. At 27°C, nodule formation, nodule development, and nitrogen fixation in soybeans were found to be most rapid (Dart et al., 1975). Nodulation was slow, initial nitrogenase activity was low, and leaves remained yellow for 30 days after planting when the day temperature of the soil was 21°C. Nodules formed at 33°C but at a slower rate than at 27°C. Nitrogen fixation was detected 19 days after planting at 27°C, but 26 and 22 days were required for measurable fixation at 21° and 33°C, respectively. Differences were observed between *Rhizobium japonicum* strains in rates of nitrogen fixation at different temperatures. Some strains were ineffective at 33°C. Greater effectiveness of certain strains was believed to be due to the enhanced nodule tissue production at higher temperatures. Nodule efficiency (nitrogenase activity per unit weight of nodule) reached a maximum between 16 and 26 days after planting for the different strains at all temperatures. Increasing temperatures from 10° to 30°C changed the frequency of the serological groups in the nodules.

A detailed study of soil temperatures under soybeans in Rhodesia revealed that the highest soil temperatures reached during the season were lethal to the rhizobia (Willatt, 1966). Shading by the plant canopy reduced the high soil temperature as much as 17°C at a depth of 2.5-cm. Planting dates affected the soil temperatures since different degrees of shading occurred during the period of highest temperatures.

In the Philippines, a mulch treatment on upland soils resulted in doubling the soybean yield of 20 cultivars. High soil temperatures probably contributed to the lower yields under unmulched conditions. Cultivars which produced the highest yields in mulched conditions were different from those producing highest yields in unmulched conditions, suggesting high temperature tolerance by some cultivars.

C. Effect of Temperature on Physiological Processes

The optimum air temperature for photosynthesis is 25°–30°C. However, optimum temperature for photosynthesis in leaves of the cultivar LEE is 35°C (Hofstra and Hesketh, 1969). In the absence of oxygen, the maximum temperature for photosynthesis is 40°C. The CO_2 assimilation by soybean canopies is reduced 20% when the canopy temperature is increased from 30° to 40°C. Photosynthesis decreases as the solar altitude declines because of decreased radiation and lower temperatures. Apparent saturation levels are related to temperature because radiation saturation generally occurs when temperatures exceed the optimum.

The rate of translocation within the plant is also affected by temperature. A reduction in temperature of the stem results in a slower rate of translocation. Translocation stops at temperatures of 2°–3°C.

Optimum temperatures also exist for nutrient uptake. Potassium content of soybeans increases with a rise in temperature up to 32°C. Zinc uptake is greatest at 30°C. In one study, when temperatures increased from 14° to

26°C, the content of zinc increased from 17 to 30 ppm in the tops of young soybean plants.

A phytotron study has shown that nitrogen content can increase 20% in plant tissue and seeds of uninoculated plants when the soil temperature is increased from 19° to 30°C. Grain yield per container was increased by 32% when the soil temperature increased from 19° to 30°C. Inoculated plants in the same study produced highest yields at 24°C. The optimum temperature for nitrogen absorption from the soil appears to be higher than that for nitrogen fixation. High soil temperatures during early and midseason may reduce nitrogen fixation, but the effect on the soybean plant may be masked by nitrogen absorption from the soil.

The respiration rate of soybean seed during the final stages of development is affected by the moisture content of the seed and, to a lesser degree, by the temperature. Seed weight losses of 0.03, 0.04, and 0.05% per hour have been observed at temperatures of 21°, 29°, and 32°C, respectively, due to respiration of high moisture seed. The respiration rate increases at higher temperatures causing a greater loss in seed weight.

D. Effect of Temperature on Seed Quality

Warm temperatures and high relative humidity during the postmaturation, preharvest period are unfavorable to soybean seed quality. The detrimental effects of alternate wetting and drying of the seeds in the pod are accentuated by high temperature. In temperate regions, early maturing cultivars often ripen while temperatures and relative humidity are high enough to promote rapid seed deterioration. For this reason, late maturing cultivars frequently produce better quality seed than do early maturing cultivars, even though relative humidity may remain high throughout the harvest season. Satisfactory seed quality in the humid tropics may depend on choosing a date of planting such that maturation will occur at the end of a rainy period. Hot, dry temperatures during seed maturation may also be detrimental to seed quality. Green and Pinnell (1968) associated reduced seed quality, as measured in laboratory germination tests, field emergence counts, and visual ratings, to such conditions.

Temperature is an important factor during the storage of soybean seed. At 40°C the soybean cotyledons deteriorate rapidly, but the epicotyl and hypocotyl remain viable. The respiration rate of soybean seed increases with temperatures up to 50°C.

Seed storage temperatures can increase from 22° to 47°C in 21 days when the moisture content of the seed is 17.6% and air exchange is prevented. With a higher water content, the storage temperature can rise more rapidly. Molds and nonbiological oxidation are primarily responsible for temperature increases during storage. However, molds are unable to grow in soybean seed having a moisture content of 12.5%. Under normal storage conditions, when the moisture level is at a level satisfactory for seed storage, the heat generated by seed respiration is insignificant.

Seed viability decreases rapidly under conditions of high temperature and high seed moisture. At 30°C and a seed moisture content of 15%, soybean seed will lose all viability in approximately 4 months. To maintain germination of soybean seeds in the tropics, it is frequently necessary to control storage temperature and/or relative humidity. Storage at a temperature of 20°C and a relative humidity of 60% should maintain soybean seed quality for 8 to 9 months, provided the seeds are of good quality when placed in storage.

E. Effect of Temperature on Protein and Oil

Soybeans usually have a higher oil content when they are grown in warm environments. Fourteen cultivars of United States origin were found to average 1.9% more oil in the tropics than when grown in their area of adaptation in the United States, but the average protein content did not differ. When temperature and oil content were correlated, the closest relationship occurred between 20 to 40 days before maturity. High temperature during the growing season was correlated with high oil content. Low temperatures during the period from flowering to maturity were associated with a high iodine number of the oil. Cultivar differences were observed for both associations of temperature with oil content and iodine number of the oil.

In tropical areas, temperatures remain relatively constant throughout the growing season, whereas in temperate areas temperatures reach a maximum during midseason and decrease towards maturity. In greenhouse studies, temperatures of 21°, 25°, and 29°C during the pod filling stage produced soybeans with 19.5%, 20.8%, and 23.2% oil. Both linolenic and linoleic acids are negatively correlated with temperature. Linolenic acid is more closely correlated with temperature than is linoleic acid.

Date of planting studies have shown that the oil content is lower when the seeds mature at cool temperatures. Early cultivars in temperate regions may escape cool temperatures during maturation and, consequently, may have a higher oil content than later maturing cultivars. The interaction between cultivar and planting date on oil content is a function of temperature during the pod filling stage of development.

Protein content in soybean seeds is usually inversely related to oil content. However, temperature does not appear to be strongly associated with protein content and has little effect on the amount found in seed. In greenhouse studies, day temperatures affected the methionine content of the protein in

4. Agronomic Characteristics and Environmental Stress 95

soybeans (Krober, 1956). At 32°C the methionine content of LINCOLN soybeans was 1.40%, but at 21°C the methionine content was 1.13%. These results suggest that soybeans produced in warm environments may have better quality protein for food use.

F. Effect of Temperature on Pests

Disease organisms have different temperature regimes. Phytophthora rot, [*Phytophthora megasperma* (Drechs.) var. *sojae* A. A. Hildebrand] a fungal root and stem disease, develops most rapidly at temperatures of 25°C and above. Postemergence mortality caused by this fungus is highest in the temperature range of 30°–35°C. Damage to the soybean roots by the rootknot nematode (*Meloidogyne hapla*) is severe at 20°C soil temperature but increases with higher temperatures. This organism produces the greatest number of galls when the soil temperature ranges from 25° to 30°C. Pythium rot (*Pythium* spp.) is a fungal disease which attacks seed and seedlings. It causes seed decay and preemergence damping-off. Different species have different temperature optima. Temperatures from 24° to 36°C are favorable for infection by *Pythium aphanidermatum*, but *P. debaryanum* and *P. ultimum* are more virulent at temperatures of 15°–20°C. The number of viable spores of brown stem rot [*Cephalosporium gregatum*, (Allington and Chamberlain)] is reduced when subjected to alternating freezing and thawing temperatures. Normal germination of the spores occurs with storage temperatures of 0°–4°C. When stored at 23°C, only 2% of the dry spores germinate. The optimum temperature for infection by soybean rust [*Phakopsora pachyrhizi*, (Sydow)] is 18°–21°C (Bromfield, 1976). Rust development is slowed when temperatures are above 30°C or below 20°C. Soybean mosaic symptoms are temperature dependent. At 18°C infected plants have more severe rugosity or crinkling of the leaves. The symptoms are almost completely masked when the infected plants are grown at 30°C or higher temperatures. There are differences in severity of symptoms due to interactions among cultivars, viral strains, and temperatures. Forms of Fusarium wilt and root rot (*Fusarium* spp.) differ in their optimum temperature regime. Some can infect at low temperatures while others cause infection only at temperatures above 28°C. Charcoal rot [*Macrophomina phaseolina* (Tassi) Goid.], another fungal disease of the soybean root and lower stem, attacks young plants when they are under stress of high temperature, low moisture or other environmental extremes. Damage by diseases, insects, and machines restricts the ability of the soybean root to supply the plant with sufficient water when temperatures are high and soil moisture is limited. In temperate regions, the low temperatures of the winter season limit the multiplication and spread of many soybean diseases. In contrast, tropical environments with warm temperatures and often humid conditions permit rapid, continuous growth and multiplication of many causal agents.

Species of *Aspergillus* invade soybeans during storage and cause deterioration in seed quality. *Aspergillus flavus* can be found seed-borne in most soybean lots (Dhingra *et al.*, 1973), although detection of the organism during bioassay is temperature dependent. When seeds are incubated at 30° or 35°C, the fungus is readily recovered; at 20° or 25°C, recovery is negligible.

Invasion of soybean seed by *Aspergillus glaucus* was found by Dorworth and Christensen (1968) to increase with increasing seed moisture content, increasing temperature, and increasing time of storage. However, invasion of low moisture seed (12.1%) or those stored at a low temperature (15°C) had no detectable effect on seed germination.

Temperature may be the single most important factor affecting insect activity. Hatching of eggs, pupae development, oxygen consumption, food consumption, and rates of metabolism are examples of insect activity which generally increase as temperature approaches an optimum. The optimum temperature may not be the same for each process. A twenty-five fold increase in the rate of adult insect population growth can result when temperature increases from 15° to 25°C. Such an increase in population is possible because of the increase in the rate of oviposition and the decrease in the time required for development from an egg to an adult (Bursell, 1970). Temperature also influences food requirements. An insect may have enough fat reserves for 5 days if the temperature is 20°C, but only enough for 2 days if the temperature is 30°C. At the higher temperature, the insect would increase its feeding rate 2.5 times. Temperature changes to extremely high or low levels may increase the death rate of insects.

The green cloverworm, *Plathypena scabra* (Fabricius), is a major insect pest of soybeans in Iowa. The adult is greatly affected by night temperature. Its flight activity is greatest when minimum nightly temperatures are 21°C or greater. The painted lady butterfly, *Cynthia cardui* (L.), a migratory species which overwinters in Mexico or southwestern United States, is another important insect pest in Iowa. Eggs do not develop at temperatures of 13°C or less. The temperature range from 21° to 27°C is optimum for the painted lady butterfly. A higher body temperature is required for flight of many insects and they may expose their wings to radiant energy or flutter their wings to generate the necessary heat before takeoff. Insects are also known to migrate when soil temperatures reach a maximum. In general, the upper limits for survival are 40°–45°C, with considerable variation among species. Death of some insects may occur at temperatures well above freezing, but a few insects can withstand complete freezing. Mite (*Tetranychus turkestani*) populations tend to increase on soybeans under hot, dry environments. The ability of the adult Mexican bean bettle [*Epilachna varivestis*

(Mulsant)] to survive freezing temperatures is positively associated with its age. The eggs hatch and pupae develop at temperatures of 12°–31°C, but temperatures near 0°C for 12 days are fatal to the pupae (Auclair, 1959).

Soybeans are subject to attack by numerous species of insects which feed on different parts of the plant. Some will attack the germinating seed, roots, and nodules in the soil. Other insects will feed on the stems, leaves, flowers, pods, and seeds, depending on their preference and growth stage of the plant. Leaf feeders may attack at different times during the season. The amount of damage caused by infestations of insects depends on the rate of development of the plant and the population of the pest. A slight amount of defoliation may actually benefit the plant by permitting increased light penetration to lower leaves in the canopy. Excessive defoliation may reduce the leaf area index below the optimum and reduce the yield potential of the soybean crop.

Temperature fluctuations from season to season in temperate regions may protect these areas from a major buildup of soybean insect pests. Without lethal temperatures to control insect populations in tropical regions, other control measures will be necessary to protect the soybean crop. In new areas of production, initial insect populations may be inadequate to cause economic damage, but population levels can be expected to increase as the area of production increases.

IV. WATER

The process of germination is initiated with the imbibition of water. For germination to proceed, the soybean seed must be able to achieve a moisture content of about 50%. This moisture content can be attained in 5 days from a soil with a moisture tension of not less than −6.6 bars (Hunter and Erickson, 1952). Soybean seed can survive for a month or more in a soil having insufficient moisture to allow swelling. Because many fungi can develop on seeds having moisture contents of less than 50%, seeds which imbibe some water but which are unable to obtain sufficient moisture for germination may perish due to fungal invasion. Soybean seed placed originally in an inadequate moisture environment (−8.0 bars) for 8 days, then transferred to an environment with sufficient moisture, failed to germinate due to the action of fungi (Hunter and Erickson, 1952).

Small soybean seed may germinate in a drier soil than will large seed. Near-isogenic soybean lines having 100-seed weights of 9.5, 13.6, and 22.6 g were planted in clay soil at moisture levels of 20, 22.5, 25, 27.5, and 30% (Edwards and Hartwig, 1971). No emergence was observed at 20% soil moisture. For each moisture level where germination occurred, however, the small and medium sized seed gave more rapid emergence and greater root development than the large seed.

The capacity of soybeans to emerge through a crust is improved when adequate moisture is available to the seed. Soil moisture content and soil crust strength were varied independently by Hanks and Thorp (1957). They found that as crust strength was increased, seedling emergence was depressed less when the soil was at field capacity than when only 25% of available moisture remained.

Water consumption by soybeans will vary with climatic conditions, management practices, and length of the growing season. Water use estimates of 64–76 cm reported by Henderson and Miller (1973) for the desert areas of southern California probably approach a maximum. In Texas, soybeans which were irrigated when soil moisture in the 0–60 cm depth was depleted to 60% available were reported to use 65 cm of water during the season (Dusek et al., 1971).

A. Effect of Mulching

Smaller-seeded soybean types emerge from a crusted soil more easily than larger-seeded types (Cartter and Hopper, 1942). Where surface crusting is a problem, the use of organic mulches may improve emergence. Mulching also increases moisture availability in the seed zone. Maintenance of plant residues on the soil surface increases infiltration of water, decreases water and wind erosion, and conserves soil moisture by reducing evaporation. The use of a grass mulch increased emergence of low quality soybean seed (Dadson and Boakye-Boateng, 1975). The beneficial effects of mulching were attributed primarily to moisture conservation and reduced crusting. More rapid and greater final emergence of soybeans were reported by Metha and Prihar (1973) with the use of a straw mulch as compared to a bare soil. The effect was particularly marked when a simulated rain was applied one day after planting. In the absence of rainfall, the beneficial effect of straw mulch was apparently associated with reduced maximum soil temperature in the seed zone. Each ton of straw applied per hectare reduced the maximum soil temperature by approximately 1°C. Following rainfall, the mulch reduced crust formation as well as soil temperature.

Soil moisture conservation can also be affected by using gravel and sand as mulching materials. Surface stones reduce erosion and evaporation, and increase soil moisture (Lamb and Chapman, 1943). Farmers in a low rainfall region in China used a pebble mulch to increase yields of high value crops (Tsiang, 1948). In contrast to plant residue mulches, gravel mulch increases maximum soil temperature (Fairbourn, 1973). In areas of high insolation, an increase in soil temperature may be detrimental to soybean germination and subsequent growth and development.

B. Effect of Water Deficit

In drying soils, emerging seedlings are somewhat protected from desiccation as the result of osmotic adjustment of the hypocotyls. This adjustment appears to depend on the addition of solutes of cotyledonary origin to the elongating cells of the hypocotyls and results in the maintenance of an almost constant cell turgor pressure under conditions of tissue desiccation. Slow but continuing growth (10% of the rate of well watered seedlings) has been observed at tissue water potentials as low as −9 bars (Meyer and Boyer, 1972).

Moisture deficiency during the vegetative stages of soybean development reduces the rate of plant growth. Leaf enlargement decreases rapidly as leaf water potential increases. Boyer (1970) found that a relatively low stress level of −4 bars reduces leaf enlargement to only 25% of the observed maximum. Photosynthesis was much less sensitive to moisture stress than was leaf enlargement, however, Ghorashy et al. (1971) found leaf water potential to be high enough to limit photosynthesis on days with high insolation. When moisture stress is sufficient to reduce photosynthesis, nitrogen fixation is also reduced, since photosynthesis, transpiration, and nitrogen fixation are correlated at leaf water potentials between −5.4 and −27.6 bars (Huang et al., 1975). The effects of moisture stress on vegetative growth are reflected in smaller leaves, reduced stem diameter, and reduced plant height.

Soybeans develop an extensive root system. The inherent rooting capacity of the soybean plant may be similar to that of cowpeas and grain sorghum (Henderson and Miller, 1973). Moisture extraction from the depth of 150–180 cm was reported in a soil without barriers to root penetration. Yield responses to early irrigation are therefore likely to be small if soil moisture conditions at planting are favorable and no impediments to rooting exist.

With limited irrigation water available, early maturing cultivars may perform better than late maturing cultivars in dry regions. Matson (1964) found an early cultivar to be less responsive to irrigation than a late one. The use of an early cultivar provides some protection against complete failure due to drought, but at the expense of maximum yield should moisture supply during the season be favorable. The effect of moisture stress on yield, however, varies considerably among cultivars.

Yield of soybeans is most affected by moisture stress during the pod filling period (Dusek et al., 1971; Doss et al., 1974). Irrigation beginning at flowering is as effective in increasing yield as is irrigation throughout the growing season (Spooner, 1961; Matson, 1964). When rainfall was prevented from reaching the soil, Doss *et al.* (1974) found that water had to be supplied throughout the entire season to obtain highest yields. Stress during any part of the season reduced yields, but greatest reductions occurred when moisture was limiting during the pod-filling stage.

Moisture stress during flowering increases abortion of flowers and young pods. When stress, from flowering through pod set, was followed by adequate irrigation during pod filling, Dusek *et al.* (1971) found a yield reduction due to a decrease in the number of pods per plant. Number of seeds per pod and weight per seed were unaffected by moisture stress during this period. Stress of short duration during early flowering usually causes little reduction in the number of pods per plant since soybeans flower over a relative long period of time. Soybeans can compensate for early flowering and pod abortion by increased set of later flowers, providing sufficient moisture becomes available (Pendleton and Hartwig, 1973). Stress during the pod-filling period reduces seed size (Whitt, 1954; Dusek *et al.*, 1971). The reduction of dry matter accumulation in the seeds with late season moisture stress may be a result of premature loss of leaf area and a shortening of the pod-filling period.

C. Effect of Excess Water

Excessive soil moisture severely restricts germination and early growth of soybeans. These effects are apparently the result of restricted oxygen movement to the seed and plant roots (Ohmura and Howell, 1960). Soil moisture per se has an effect on germination of soybeans independent of those on soil aeration. At soil moisture tensions of −0.3 bars or more, Grable and Danielson (1965) found that pathogenic organisms developed rapidly on seeds and roots of germinating soybeans and almost completely stopped all root growth.

On soils which pond water, surface drainage may be effective in improving germination. On a poorly drained field characterized by long periods with ponded water and formation of dense crusts upon drying, Fausey and Schwab (1969) reported sparse stand establishment and essentially no growth. Effective drainage resulted in increased plant stand, root development, plant height, and yield.

Placement of dry soybean seed in wet soil may result in reduced germination, growth, and yield of soybeans. A dry seed placed in wet soil imbibes water rapidly. Injury may occur as the result of more rapid hydration of outer cotyledonary tissue in comparison with the inner tissue, thus causing a tension crack between wet and dry tissue of the cotyledon (Sorrells and Pappelis, 1976). Imbibitional injury is most severe when dry seeds imbibe at low temperatures, but Obendorf and Hobbs (1970) found decreased survival and early growth when seeds with an initial moisture content of 6% were imbibed at 26°C as compared to seeds with an initial moisture content of 16%. Cultivars were not all equally affected by imbibition at the low initial moisture content.

Once soybeans are established, the effect of an excess or deficiency of moisture will depend on the stage of plant growth at which stress occurs and its duration. Yield reductions in two cultivars as the result of extended flooding were reported by Spooner (1961). The cultivars differed markedly in their sensitivity to flooding. Over a 3-year period, yield of the cultivar DORMAN was reduced an average of 8, 38, and 59% by flooding at the time of flower initiation for 7, 14, and 21 days, respectively, whereas yield of the cultivar LEE was unaffected by 7 days of flooding and reduced only 6 and 18% by 14 and 21 days of flooding, respectively. Soybeans were flooded for either 15 or 30 days during the vegetative and reproductive stages of the crop by Barni and Costa (1976a,b). Flooding for 15- and 30-day duration prior to initiation of flowering reduced plant height by 21 and 32%, respectively. Comparable treatments reduced height of the first pod by 25 and 30%. Thirty days of flooding, beginning at the initiation of flowering, hastened maturation by 5 days whereas a shorter period of flooding at the same growth stage, or flooding at later developmental stages, resulted in approximately an 11-day delay in maturation. As observed by other researchers (Gupta and Hittle, 1970), flooding at or before the pod-filling stage resulted in the formation of aerenchymatic tissues and adventitious roots. The capacity of a crop to produce adventitious roots is associated with its ability to tolerate flooding (Kramer, 1951).

Yield was reduced by flooding at all times except near the end of the growth cycle (Barni and Costa, 1976a). Greatest decreases resulted from 30 days of flooding during flowering, prior to flowering, and at the beginning of pod fill, in that order. Decreases were 66, 40, and 28%, respectively, when compared to the nonflooded control which yielded 4088 kg/ha.

Prolonged flooding generally increases oil percentage in seed and decreases protein. In research conducted by Barni and Costa (1976b), leaves became chlorotic several days after initiation of flooding, suggesting disruption of nitrogen fixation. No nitrogen fixation (as measured by acetylene reduction) was detected in flooded soil by Huang et al. (1975). Drying the soil eliminated the inhibitory effects of flooding, suggesting that inhibition of nitrogen fixation results from restricted gas exchange between the nodules and the atmosphere. Inhibition of nitrogen fixation at subambient levels of oxygen is completely reversible and soybeans eventually regain their normal green color even after 30 days of flooding.

Poor quality seed was produced by soybeans subjected to flooding (Barni and Costa, 1976a). Flooding for 15 or 30 days at any time after the initiation of flowering resulted in seed with less than 50% germination.

D. Effect of Atmospheric Humidity

Yield of soybeans grown with adequate supplies of soil moisture can be affected by atmospheric humidity. A 21% reduction in yield was recorded for soybeans grown at day/night relative humidities of 47/46% as compared to 81/84% (Woodward and Begg, 1976). Soybean yield decreased at the low atmospheric humidity as a result of a reduction in pod number and thus bean number, which was only partially compensated by a small increase in weight per seed. The reduction in pod number was associated with flower abortion. The total dry weight of plant tops, dry weight of stems, and the number of nodes per plant were reduced in the low humidity environment. Internal moisture stress may have been induced by high evaporative demand even though moisture in the soil was adequate. An additional consequence of low humidity is increased shattering. Many cultivars developed in environments with high relative humidity during the harvest season will dehisce as soon as mature when grown in arid zones. Considerable variation is present, however, for this characteristic, and cultivars such as LEE have considerable resistance to shattering.

Water is essential for plant growth, but too much or too little moisture at any given stage of growth may be detrimental. Management practices, such as row spacing, mulching, and minimum tillage help to conserve water which may be limiting. Excess water may be more difficult to manage, but the use of drainage tiles, terraces, dams, and other water management practices help to reduce crop damage caused by too much water.

Consistently high soybean yields may be difficult to obtain in arid regions of the world because of low atmospheric humidity and its effect on photosynthesis. Even when irrigation water is available, the internal plant structure may limit the supply of water to the leaves and reduce the growth rate of the plant. Seed shattering is also a factor to be considered before growing soybeans in low humidity environments.

V. WIND

A. Effect of Wind

Wind is a major environmental factor directly or indirectly influencing the productivity of a soybean crop. Wind increases the rate at which the soybean canopy loses or gains heat by convection. The rate of heat exchange by this process is related to wind speed, but the relationship is not linear; most of the increase in rate of convective cooling is accomplished by a wind of only 3 to 5 km per hour (Gates, 1965). Wind movement also assists to replenish CO_2 fixed by the crop during daylight hours.

Wind greatly increases potential evapotranspiration. In arid climates, heat advection into well-irrigated areas induces extremely high rates of evapotranspiration (Fritschen, 1966). Thus, with the aid of winds, energy available for evapotranspiration can exceed the energy of net radiation. When potential evapotranspiration is high, soil-water potential must be maintained at a high level in order to prevent moisture stress. Light stress may also occur when wind prevents the soybean plant from orienting its leaves toward the sun. Radke and Burrows (1970) suggest that wind increases exposure of the more reflective underside of soybean leaves to light and thereby decreases efficiency of light utilization. Water-stressed leaves have less ability to maintain a normal orientation than do leaves not stressed for water (Radke and Hagstrom, 1973).

The velocity, frequency, duration, and direction of wind interact and may cause mechanical and abrasive damage to soybeans. Leaf area may be reduced by wind, although this is rare unless the wind is accompanied by hail or sand. Hail can both shred leaves and bruise or break stems. The abrasive action of windblown sand can also destroy soybean tissue. Plants weakened by disease and/or insect damage can easily be lodged by wind. Wind direction is a factor in lodging. When the wind direction is parallel to the row, fewer plants lodge at the same wind speed than when wind direction is perpendicular to the row. Adjacent plants tend to support one another within the row and less force is applied to an individual plant. Yield losses can be severe when lodging occurs early and in a productive environment (Cooper, 1971).

Wind aids in the distribution of plant pests. Wind direction and velocity affect insect distribution in several ways. Long distance spring migrations of insects from warmer equatorial climates to the harsher, more temperate zones is mainly accomplished by low-level jet winds. These winds bring insect pests such as the potato leafhopper [*Empoasca fabae* (Harris)] and thrips [*Sericothrips variabilis* (Beach)] into soybeans in the midwestern United States. Lewis (1964) suggested that a slight convection and unsettled weather were important ingredients in mass flights of several groups of thrips. Field-to-field distribution of insect pests is also often controlled by wind factors. Haines (1955) stated that winds above 11 km/h delayed take-off of winged aphids. This has also been shown for thrips (Lewis, 1964) and other insects. Within fields, movement of insects is similarly influenced. In recent studies conducted at Urbana, Illinois, within-field aphid movement was directly related to wind velocity and direction. Resulting virus spread, a good indicator of vector movement, was downwind of the source area (R. M.

Goodman and M. E. Irwin, unpublished). Wind is so important in understanding pest outbreaks that they cannot be elucidated without considering the causal relationships between redistribution of insect populations and the nature of the windfields which transport and concentrate them (Joyce, 1976).

Microscopic spores and light-weight weed seeds may also be transported by wind. It is suspected that spores of soybean rust [*Phakopsora pachyrhizi* (Sydow)] are carried from tropical to temperate regions in Southeast Asia each year by wind movement. Spread of spores of many pathogenic organisms over short distances is enhanced by wind.

B. Effect of Windbreaks

Windbreaks often markedly increase growth and yield of soybeans. When moisture was adequate for plant growth, Frank *et al.* (1974) found that soybeans sheltered by a corn windbreak grew taller, produced more leaf area and dry matter, and yielded 22.5% more than unsheltered plants. Radke and Burrows (1970) found that sheltered and unsheltered soybeans used similar quantities of water, but that water use efficiency of sheltered soybeans was greater due to higher yields. Under conditions of limited moisture, the use of windbreaks may be less beneficial. Sheltered plants grow more rapidly than do unsheltered plants and may exhaust the available soil moisture before producing a crop. Frank *et al.* (1974) found no difference in yield between sheltered and unsheltered soybeans under conditions of limited soil moisture. Canopy temperature of the sheltered plants was 2.4°C higher than that of the unsheltered plants.

Windbreaks reduce windspeed and potential evaporation from sheltered areas. Tall wheatgrass [*Agropyron elongatum* (Host) Beauv.] barriers have been shown to reduce evaporation from the soil surface and improve the soil water environment for seed germination and stand establishment (Aase and Siddoway, 1976). Transpiration from the sheltered crop canopy may likewise be reduced and photosynthesis may benefit from an increase in stomatal conductance. The potential beneficial effects of a windbreak are not consistent over the whole width of a sheltered strip. Radke and Hagstrom (1973) found that windspread and potential evaporation were reduced for only the first 7 or 8 of 14 sheltered soybean rows. Higher potential evaporation rates occurred in rows 8 to 14 than in unsheltered soybeans and air temperature was 1.5°C higher. Also, because of competition, yield of soybeans nearest a corn windbreak may show no yield advantage (Radke and Burrows, 1970). Because of this competition, a minimum windbreak spacing of six times the height of the corn was suggested. However, optimum windbreak spacing will depend on environmental factors and the type of windbreak used.

In areas where winds create a high evaporative demand or cause frequent plant damage, windbreaks may prove to be useful. Small increases in air temperature and relative humidity are frequently experienced behind windbreaks. Usually, the total water used by the sheltered crop is not reduced, but it is used more efficiently. Wind erosion is diminished. Vegetative growth and final yield of soybeans is often increased. Their inconvenience has been a deterrent to use of windbreaks in large-scale, mechanized agriculture.

VI. PESTS

A. Birds and Rodents

Damage to soybeans by birds and rodents usually occurs during germination of the seed or during the seedling stage. The soybean seed and cotyledons are an acceptable food and attract pests when alternative food supplies are limited. Damage by either birds or rodents may be severe if the area planted is small and other desirable feeding areas do not exist. Reports of bird or rodent damage are infrequent in major soybean production areas such as the Mississippi Valley in the United States.

The eared dove (*Zenaida auriculata*) causes damage to 75% of the soybean fields in the Cauca Valley of Colombia. Because of the larger number of feeders, damage is greatest when the time of planting coincides with the time of fledge for the young birds. Delayed planting reduces the problem. Abnormal environmental conditions, such as heavy rainfall or drought, may force the birds from traditional feeding areas into the cultivated fields for food. Nesting may occur in wooded areas a considerable distance from the fields if cultivated areas are extensive. The eared dove has been known to migrate 20 km daily to feed in soybean fields.

An examination of dove populations in Colombia revealed that about 14% of the birds had soybean seed and 2% had soybean cotyledons in their guts. Individual birds had consumed as many as 55 soybean seeds. Examination of the feeding area exposed piles of up to 30 cotyledons which had been regurgitated by doves after overfeeding. Death of the bird may occur when moisture is absorbed by a large number of seeds, causing expansion of the cotyledons and thereby suffocation of the bird.

Significant soybean yield losses may be caused by bird damage. Most of the damage will result from feeding prior to emergence of the seedlings. The reduction in plant stand affects the development of the remaining plants. Removal of the cotyledons of seedlings may reduce plant development if the unifoliolate leaves have not emerged. Less serious damage is caused if only portions of the cotyledons, or if only one cotyledon, are removed before leaf development.

Rodents such as rats (*Rattus rattus*) are known to eat soybean seed and seedling parts resulting in damage similar to that caused by birds. The amount of damage depends on the alternate food supplies, size of the soybean area, location of hiding areas, and environmental factors which affect the rodent population or feeding habits. Yield was decreased significantly when both cotyledons were removed during the first 2 days after emergence in a simulated rat study in Colombia. A reduced plant population is the main result of seed or seedling damage by most pests. When the damage reduces the plant stand to less than optimum, the yield potential may also be reduced. Little or no apparent damage is caused by rodents or birds after the seedling stage. The fibrous pod serves as a good storage package for the seed during development and maturation.

B. Diseases

Numerous diseases which affect the soybean plant have been identified. Leaf diseases reduce the leaf area or the photosynthetic potential of the plant. Root and stem diseases usually restrict the absorption and transport of nutrients and water by the plant. Lodging of plants is often due to root or stem weaknesses related to diseases. Pod and seed diseases affect the quality of seed produced. Either directly or indirectly, most diseases of soybeans have a yield depressing effect. The severity of damage varies among diseases and the interaction of the causal agent with the environment. Overall, diseases cause an estimated 10 to 15% yield reduction annually.

One of the most important diseases in the soybean production area of the United States is brown stem rot [*Phialophora gregata* (Allington and Chamberl.) W. Gams.]. Resistance to brown stem rot is not present in most soybean cultivars and crop rotation is the most common control practice used. The fungus develops more rapidly in plants with advanced maturity or during periods of cool temperature. With browning of the vascular system the rate of water flow through the stem to the pods and leaves is reduced. Rather than a physical plugging of the vessels by fungus mycelium a metabolite of the fungus causes the restriction of water flow through the stem (Chamberlain, 1961). The water flow may be reduced to 10% of the normal rate. Such severe restrictions are likely to result in desiccation and death of leaves in the upper portion of the plant during periods of intense transpiration. Premature ripening accompanies the disease and results in smaller seed size, especially on the upper nodes. Late maturing cultivars are usually more susceptible to damage than early maturing cultivars. A reduction in seed number accounts for most of the yield reduction which has been reported to be as much as 25%. Lodging of infected plants results in slower harvesting and greater harvest losses.

4. Agronomic Characteristics and Environmental Stress

Phytophthora rot, caused by the fungus *Phytophthora megasperma* (Drechs) var. *sojae* A. A. Hildebrand, may cause damage to soybeans at any stage of development. The disease has only been reported in North America. Phytophthora rot infection depends on the environment and cultivar susceptibility. The most severe damage occurs on low, poorly drained, clay soils. Young soybean plants are most susceptible and may die. In poorly drained soils, the seed may rot and preemergence damping-off will reduce plant stands. Wilting and death of seedlings may result when the root and stem become infected after emergence. When infection occurs after the leaves have developed, the leaves turn yellow and wilt. Older plants may also wilt and die, but more gradually, beginning with the lower leaves and moving upward. Infected branch roots are almost completely destroyed and infected taproots become discolored and less efficient. The discoloration may progress up the stem to the lower branches. Yield losses of more than 50% have been attributed to phytophthora rot. Seed produced by damaged plants is reported to have lower protein and higher oil content than seed from resistant plants.

Bacterial pustule [*Xanthomonas phaseoli* (E. F. Smith) Dowson var. *sojensis* (Hedges) Starr and Burkh.] is a leaf disease of soybeans found in most parts of the world where warm, moist conditions exist. Most improved cultivars carry resistance to the disease. Bacterial pustule symptoms include chlorophyll and leaf tissue destruction which reduce the photosynthetic potential of the plant. Both seed number and seed size are reduced because of premature defoliation. The degree of infection is influenced by environmental conditions. Yield varies inversely with the degree of infection. Losses up to 15% have been reported.

A fungal leaf disease found primarily in the Eastern Hemisphere is soybean rust [*Phakopsora pachyrhizi* (Sydow)]. Recent reports indicate its presence in the Western Hemisphere. None of the cultivars commonly grown in the United States are resistant to the fungus. Disease tolerance has been reported in a few experimental soybean lines and the tolerance is being transferred to improved cultivars. Chlorotic or brown spots appear on the leaf surfaces and spread to the petioles and smaller stem parts. Pustules develop in these spots and reduce the photosynthetic area of the plant. Rust may cause premature defoliation and a reduction in seed weight and seed number. Yield reductions from 20 to 50% are commonly reported in Southeast Asia; complete loss of the soybean crop occurs in some cases. Warm temperatures and frequent rainfall result in heavy damage, but temperatures above 30°C or heavy rainfall tend to decrease sporulation of the fungus.

Soybean mosaic, caused by the soybean mosaic virus (SMV) may be the most common viral disease of soybeans in the world. SMV may be carried in the seed or spread by aphids and mechanical means. Infected seed may fail to germinate, or produce diseased seedlings. Infected seedlings are often stunted with short internodes. The leaves are frequently rugose or crinkled and become prematurely chlorotic. Diseased pods are usually abnormal in size and shape and contain fewer seeds than pods of uninfected plants. Seeds from diseased pods are often smaller and less vigorous than seeds from healthy pods. Infected seeds may have a mottled seed coat. Nodules produced on infected plants are smaller in size, weight, and number and the leghemoglobin content is reduced when compared to healthy plants. Yield reductions of 25% or more have been reported in fields naturally infected by SMV. Plants artificially inoculated with SMV have suffered as much as 93% yield reduction.

Plant parasitic nematodes may be an important factor in some areas of soybean production. Apprxoimately 50 species of nematodes have been identified as being associated with, or pathogens of, soybeans. The soybean cyst nematodes [*Heterodera glycines* (Ichinoche)] and the root-knot nematode (*Meloidogyne* spp.) are the most common. The soybean cyst nematode has been reported in 13 soybean producing states of the United States. The root-knot nematode is primarily found in the southern half of the United States. In some areas of the southern United States other species of nematodes are more damaging than those mentioned above. The yield loss caused by all nematodes is estimated to be 10% of the annual production.

Nematodes cause primary damage to the roots and thereby affect the development of the above-ground portions of the plant. Mechanical damage to root tissue may occur when the nematodes move while feeding. Secretions injected into the root tissue while feeding may result in galls or root enlargements at the feeding site. Injury, such as root lesions, surface necrosis, devitalized root tips, the reduction of secondary roots, and root rotting may result from nematode damage. In the presence of nematode injury, nodulation by *Rhizobium japonicum* may also be reduced.

The above-ground symptoms in soybeans caused by the root-feeding nematodes are similar to other conditions which cause abnormal root systems. These symptoms include chlorosis, wilting, stunting, and possibly death of the plant. The degree of damage depends on the nematode population, soil environmental conditions, competition of other organisms, and the susceptibility of the plant to the nematode. Poor soil fertility, drought, and other environmental conditions which slow soybean plant development may accentuate the degree of damage caused by nematodes. Since nematodes and fungi cohabit the same environment, they often interact to produce greater damage than when either acts alone. The mechanical injury caused by nematodes when feeding or penetrating roots often provides pathways for infection by fungi and other soil organisms.

The above mentioned are only a few of the more than 100 diseases which are known to affect soybeans. They represent different types of causal agents which attack different parts of the plant. A given disease may cause severe damage one season and not occur the next season. These differences are due to the environmental requirements of a specific pathogen. Physical damage to plant parts by insects, man, or diseases may provide entry to additional organisms. When two or more diseases attack the same plant, a synergistic effect often results in more severe damage than either disease alone would have caused. The extent of damage caused by diseases of soybeans may never be fully realized since it is difficult to measure and quantify all types of damage to the soybean crop. More details about specific diseases are available in the "Compendium of Soybean Diseases" (Sinclair and Shurtleff, 1975).

C. Insects

Until the recent expansion of soybeans into tropical and subtropical areas, insects, except in the Orient, were not widely recognized as a serious problem in soybean production. However, the number and variety of insects associated with the soybean crop are great, and loss of yield and/or quality can result under some conditions due to insect attack.

The pattern of colonization of soybean fields by insects when the crop is introduced into a new area has been described by Turnipseed and Kogan (1976). Generally, colonization is by three main components of the native insect population. Complexes of insects such as grasshoppers, cutworms, stinkbugs, and certain caterpillars and leaf hoppers readily move into the new crop. A complex of insects with food preferences restricted to a variety of cultivated and wild legumes also readily colonize the new crop. Examples of such insects are the Mexican bean beetle [*Epilachna varivestis* (Mulsant)], which usually feeds on bean species, and the green cloverworm [*Plathypena scabra* (Fabricius)] and the alfalfa caterpillar [*Colias eurytheme* (Boisduval)], both generally associated with forage legumes in the United States. The third category of colonizers consists of certain specialized feeders that shift their host preferences, sometimes from other plant families.

Rapid accumulation of insect species in soybeans is mainly caused by colonizers of the first two categories. The third category contributes a small number of species that slowly evolve to exploit unoccupied feeding niches after the other colonizers have become established. That the second and third groups of feeders may gradually replace the general feeders is suggested by the rather restricted feeding habits of soybean insect populations in China, Korea, and Japan where several thousand years have been available for host specificity to evolve.

The species most representative of the main soybean growing zones of the

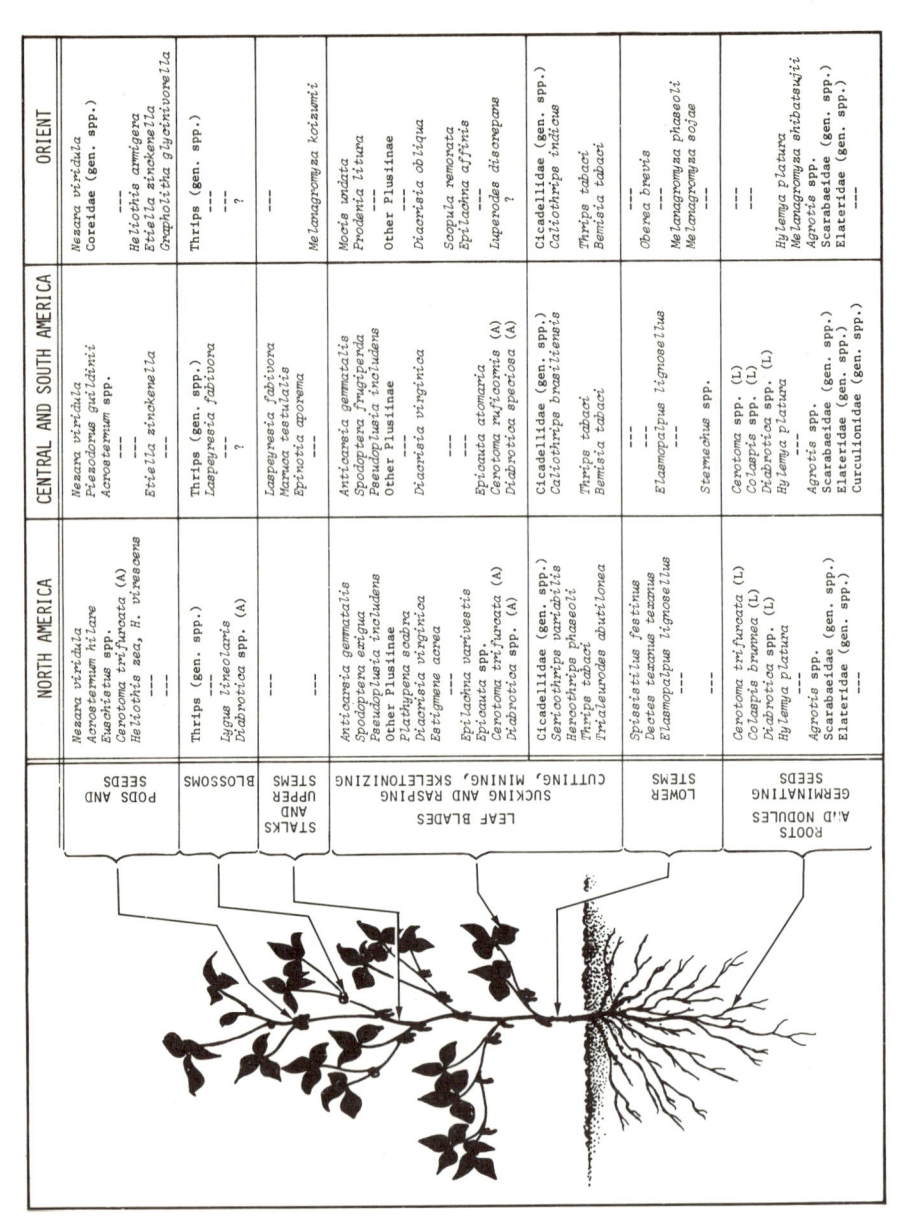

		NORTH AMERICA	CENTRAL AND SOUTH AMERICA	ORIENT
PODS AND SEEDS		*Nezara viridula* *Acrosternum hilare* *Euschistus* spp. *Cerotoma trifurcata* (A) *Heliothis zea, H. virescens*	*Nezara viridula* *Piezodorus guildinii* *Acrosternum* spp. --- *Etiella zinckenella*	*Nezara viridula* Coreidae (gen. spp.) --- *Heliothis armigera* *Etiella zinckenella* *Gnapholitha glycinivorella*
BLOSSOMS		Thrips (gen. spp.) *Lygus lineolaris* *Diabrotica* spp. (A) ---	Thrips (gen. spp.) *Laspeyresia fabivora* ? ---	Thrips (gen. spp.) --- ? ---
STALKS AND UPPER STEMS		--- ---	*Laspeyresia fabivora* *Maruca testulalis* *Epinotia aporema*	--- --- *Melanagromyza kotzumii*
LEAF BLADES	SUCKING AND RASPING	*Anticarsia gemmatalis* *Spodoptera eridua* *Pseudoplusia includens* Other Plusiinae *Plathypena scabra* *Diacrisia virginica* *Estigmene acrea*	*Anticarsia gemmatalis* *Spodoptera frugiperda* *Pseudoplusia includens* Other Plusiinae --- *Diacrisia virginica* ---	*Mocis undata* *Prodenia litura* Other Plusiinae --- --- *Diacrisia obliqua* ---
	CUTTING, MINING, SKELETONIZING	*Epilachna varivestis* *Epicauta* spp. *Cerotoma trifurcata* (A) *Diabrotica* spp. (A)	*Epicauta atomaria* *Cerotoma ruficornis* (A) *Diabrotica speciosa* (A)	*Scopula remorata* *Epilachna affinis* ? *Luperodes discrepans*
		Cicadellidae (gen. spp.) *Sericothrips variabilis* *Heracothrips phaseoli* *Thrips tabaci* *Trialeurodes abutilonea*	Cicadellidae (gen. spp.) *Caliothrips brasiliensis* --- *Thrips tabaci* *Bemisia tabaci*	Cicadellidae (gen. spp.) *Caliothrips indicus* --- *Thrips tabaci* *Bemisia tabaci*
LOWER STEMS		*Spissistilus festinus* *Dectes texanus texanus* *Elasmopalpus lignosellus* ---	--- --- *Elasmopalpus lignosellus* *Sternechus* spp.	*Oberea brevis* --- *Melanagromyza phaseoli* *Melanagromyza sojae*
ROOTS AND NODULES GERMINATING SEEDS		*Cerotoma trifurcata* (L) *Colaspis brunnea* (L) *Diabrotica* spp. (L) *Hylemya platura* *Agrotis* spp. Scarabaeidae (gen. spp.) Elateridae (gen. spp.)	*Cerotoma* spp. (L) *Colaspis* spp. (L) *Diabrotica* spp. (L) *Hylemya platura* *Agrotis* spp. Scarabaeidae (gen. spp.) Elateridae (gen. spp.) Curculionidae (gen. spp.)	*Hylemya platura* *Melanagromyza shibatsuzii* --- --- *Agrotis* spp. Scarabaeidae (gen. spp.) Elateridae (gen. spp.)

Fig. 5. Occupation of feeding niches in soybean in North America, Central and South America, and the Orient (Turnipseed and Kogan, 1976).

world are shown in Fig. 5 (Turnipseed and Kogan, 1976). The species are grouped in accordance with the part or parts of the soybean plant that they customarily attack. If different species appear in the same row, they represent possible ecological homologues. If homologues cannot be readily recognized, this is indicated by a dashed line, suggesting an unoccupied feeding niche on soybeans in the particular geographical region.

Insects may feed on soybeans from the time that they are planted until the crop is harvested and utilized. In the field, damage can result from insects feeding on any of the plant parts depicted in Fig. 5. Certain insects also provide access for disease organisms or transmit them directly to plants. Feeding may also occur on stored beans, although this is rare except under conditions of high humidity and temperature.

Numerous early season insects feed on germinating seed or seedlings. Such feeding may cause stand reduction, necessitating replanting. Studies in the United States and Brazil have demonstrated that normal stands can be reduced approximately 50% early in the season without significant losses in yield. Mechanical stand reductions of up to 45% caused no yield reduction if damage occurred before or during blooming (Caviness and Miner, 1963). Only a 15% reduction in yield occurred when 45% of the plants were removed 2 weeks after full bloom. Damage by insects attacking the lower stem of older plants may be restricted to girdling at or near soil level. Girdled plants may later lodge as the result of weakened stems. Natural girdling of 68% of the plants in a normal stand by the three-cornered alfalfa hopper [*Spissistilus festinus* (Say)] did not result in yield losses (Tugwell and Miner, 1967). H. C. Minor and N. Neumaier (unpublished) found no relationship between attack by the lesser cornstalk borer [*Elasmopalpus lignosellus* (Zeller)] and productivity of plants which survived until maturity.

Most economic losses resulting from insects occur as the result of outbreaks of foliage and pod feeders during the reproductive phases of soybean development. Yields are seldom significantly reduced by loss of from 30 to 50% of the leaf area before flowering (Kalton *et al.*, 1949; Camery and Weber, 1953; Turnipseed, 1972; Todd and Morgan, 1972; Thomas *et al.*, 1974a). Soybeans apparently compensate for such damage through the development of new leaves and by increased photosynthate production in lower leaves. After vegetative development ceases, however, the soybean plant's ability to sustain loss of leaf area without reductions in yield decreases. Turnipseed (1972) removed 17, 33, 50, and 67% of the soybean leaf area at midbloom, pod set, pod fill, and on a continuing basis from midbloom to pod set. A 17% defoliation did not cause significant yield loss at any stage of growth or on a continued basis from midbloom through pod set. Foliage losses of 33% at midbloom also did not result in significant yield losses. Yields were generally reduced following 50 to 67% defoliation at bloom, pod fill, or on a continuing basis.

Fig. 6. Severe insect damage of soybeans caused by the velvetbean caterpillar (*Anticarsia gemmatalis*) in Brazil.

Early, severe defoliation (Fig. 6) may reduce final plant height (Kalton *et al.*, 1949). Such defoliation can also result in delayed maturation, however, severe loss of foliage in the pod-filling period hastens maturation and decreases seed size. Turnipseed (1972) reported an increase in oil content and a decrease in protein content for most defoliation treatments which resulted in yield reductions.

Several workers (Hammerton, 1972; Turnipseed, 1972; Todd and Morgan, 1972) have reported a tendency for defoliation to be more detrimental to yield at higher yield levels than at lower yield levels. Therefore, as yield levels increase, degrees of damage which can be tolerated may be lower than those acceptable under current production conditions.

Certain insects feed directly on the pod. Studies which have simulated insect feeding by artificially removing or damaging pods indicate a high degree of compensation by plants when damage is inflicted during early stages of pod development. Turnipseed (1973) removed all pods over 1.2 cm long at the beginning of seed enlargement, with no significant effect on yield. Removal of pods after this stage, however, resulted in yield reductions. Pod sucking insects, of which the southern green stinkbug [*Nezara viridula* (Linnaeus)] is the most widely spread, can affect soybeans directly by reducing yield and seed quality (Daugherty *et al.*, 1964; Miner, 1966; Thomas *et al.*, 1974b; Todd and Turnipseed, 1974). Damage to the seeds generally results in lower germination percentage, decreased seedling vigor, smaller seed size, reduced oil content, slightly increased protein content, and reduced storability. However, extent of damage is a function of time of attack by the insect. A heavy infestation of stinkbugs early in the pod-fill stage will cause a more drastic reduction in yield and seed quality than a comparable infestation later in the pod development period. Several species of stinkbugs affect soybean yield and quality indirectly by transmitting the causal organism of the yeast spot disease (*Nematospora coryli* Peglion).

The soybean has substantial capacity to compensate for damage by insects, especially when this damage occurs during the vegetative stages of plant development. Yet, economic losses can and do occur. That they are not more frequent and serious is due to the presence of numerous natural enemies of insects and the use of chemical pesticides.

D. Weeds

Weeds compete with soybeans for light, nutrients, and water. Environmental conditions are generally more favorable for weed growth in the tropics and subtropics than in temperate regions. Greater yield reductions occur in environments which favor weed growth and reproduction. Soybean yield reductions of 25% have been attributed to weeds in temperate regions. In the tropics, weed control may increase yields by as much as 100%. The first 30% of the life cycle of soybeans is the most important period for controlling weeds if yield is to be maximized. In tropical regions, a 30- to 40-day weed-free period after planting is required for yields to equal those of weed-free conditions throughout the season. When plots in Nigeria were weeded at 1 and 4 weeks after emergence of the soybeans, yields were depressed 6%. Yield reductions of 12% or greater were reported when only one weeding was made. When weeds were not controlled until 10 days after emergence, a 10% reduction in yield occurred. In Brazil, reports indicate that weed competition has reduced soybean yield from 13 to 89%.

Many factors such as weather, seedling vigor, row spacing, soybean growth habit, weed population, and weed species affect the degree of yield loss caused by weeds. The optimum growing conditions for weeds are often the same as the optimum conditions for soybeans. When soil fertility, soil moisture, and temperature are favorable for soybean growth, competition from adapted weed species increases. Weed growth and soybean yield reductions are greater in fertilized plots than in unfertilized plots (Staniforth, 1962). Yield reductions are also greater when the soybean population is low. Weeds have the ability to accumulate plant nutrients at the expense of the soybean crop. Some weeds will germinate and become established sooner than soybeans when moisture is limiting. In Colombia, studies have shown that weed populations were higher and soybean yield reductions were greater during seasons with higher rainfall. In Iowa, yellow foxtail [*Setaria lutescens* (Weigel) Hubb] reduced soybean yields by 5% when soil moisture was either adequate or severely limiting. However, adequate soil moisture until late July, followed by severely limited moisture until maturity, resulted in 14% yield reductions caused by foxtail infestations. The foxtail was better able to utilize the available moisture for growth so that moisture stress during the pod-filling period resulted in reduced soybean yields.

Yellow foxtail, Pennsylvania smartweed (*Polygonum pennsylvanicum* L.) and velvet leaf [*Abutilon theophrasti* (Medic)] cause similar yield reductions. These weeds reduce plant height, increase lodging, and delay maturity slightly. Giant foxtail [*Setaria faberii* (Herrm.)] becomes a great competitor after soybeans begin to flower. A dense population of this weed reduces the light available to the soybeans. Yield reductions of 60% have been reported due to giant foxtail. Sicklepod (*Cassia obtusifolia* L.), at an average density of 7.7 plants per square meter, has been reported to reduce soybean yields by 35%. When the plots are weeded until 4 weeks after emergence, maximum yields are obtained. Morning glories [*Ipomoea purpurea* (L.) Roth and *I. hederacea* (L.) Jacq.] reduced soybean yields 52% when uncontrolled in studies at the University of Delaware. Yield reductions ranged from 12% with one morning glory every 60 cm of row to 44% with 16 morning glories every 60 cm of row. The presence of morning glories causes severe lodging and makes harvesting of soybeans more difficult. Smooth pigweed (*Amaranthus hybridus* L.) populations of one plant per meter of row can cause an 18% yield reduction. A reduction of 51% was reported when there were 40 pigweed plants per meter of row. In Colombia, purple nutsedge (*Cyperus rotundas* L.) reduced soybean yields by more than 80% (Fig. 7) and common annual weeds reduced yields by 60%. The common cocklebur (*Xanthium pennsylvanicum*) is a serious weed pest in the United States and can cause yield reductions of up to 75%. A 6% increase in soybean yield was reported for each 10% increase in cocklebur control (Anderson and McWhorter, 1976). Soybean market grades were reduced by cockleburs because of increased

Fig. 7. Soybean competition from nutsedge (*Cyperus rotundus*) soon after hand weeding in Colombia.

foreign matter in with the soybean seed. High populations of johnsongrass [*Sorghum halepense* (L.) Pers.] can reduce soybean yields as much as 42% compared to weed-free conditions. Harvesting soybeans infested with smooth pigweed and giant foxtail in Illinois, before the weeds were desiccated by frost, resulted in harvest losses which were double those for weed-free plots (Nave and Wax, 1971).

Under temperature regimes from 18° to 30°C most weed growth rates increase as temperature increases, as do growth rates for soybeans. During the first 2 weeks after emergence, soybeans normally grow faster than most weeds, but after the soybean height of 15 cm is attained, the weeds usually have the more rapid growth rate.

Weeds affect soybeans in many different ways. The plant stand can be reduced by weed competition. Shading of soybean plants by weeds during the early reproductive period may reduce pod set. Yield reduction is mainly due to a reduction in pod number (Knake and Slife, 1962). Weed competition has little or no effect on seed size or the number of seed per pod.

Weed-free soybean fields produce higher yields than those fields with weeds, providing other conditions are similar. A key to good soybean management is adequate weed control. Chemical control is usually the most efficient means of control, but may not be the most economical in all situations. Cultivation is a good weed control practice and may be used in combination with herbicides for optimum weed control. Regardless of the method used, weed control is essential to maximize soybean yields.

VII. CONCLUSIONS

The environment at any location will ultimately determine the adaptability of a soybean cultivar to that location. Soybeans are most commonly produced in environments similar to those where the soybean originated. The growth and development of soybeans from temperate regions are greatly altered when the soybeans are grown under tropical or subtropical environments. Shorter day length triggers the flowering mechanism earlier than in temperate areas, and warmer temperatures during the seedling and maturation stages speed up plant development. Therefore, it is not surprising that cultivars require fewer days from planting to harvest in the tropics.

Cultivars which require the full season to mature are usually selected for planting in the temperate regions. Yield is often less than the potential if a cultivar matures before the end of the season. Yield also suffers if the cultivar is not mature at the time of the first frost. Plant breeders have developed high yielding cultivars which are adapted to the different environments of the major soybean production regions in the United States and many environments of other countries. Seasonal weather variations, at any time during the growing season, may cause a stress that will affect an agronomic characteristic of the soybean. At a given location, the temperature, moisture, wind, and pest regimes may vary from season to season. Usually the stresses have a depressing effect on yield. Not all characters of the plant are uniformly affected by a single stress. The cumulative effect of these stresses on plant development and yield may be impossible to disentangle. Some stresses may be beneficial, but most are detrimental to the soybean plant. Not all stresses are understood, but the resultant effect on the plant is known. An example is the high percentage of flower and pod abortion in soybeans. Yield potential is reduced when flowers and pods abort, but the stress which causes the abortion is not fully understood.

The soybean plant is now grown in a wide range of environments. The FISKEBY cultivars were developed in Sweden and are some of the best adapted cultivars at 45° latitude or greater. ORBA is a cultivar recently released in Indonesia which is particularly well adapted to low altitude environments near the equator. WILLIAMS, which was developed in Illinois, is one of several cultivars which has a wide range of adaptation and produces acceptable yields in dissimilar environments. With increased emphasis on breeding soybeans which are better adapted to environments not commonly found in present major production areas, new cultivars will soon be available for production in environments where the soybean is now relatively unknown. Soybean production continues to expand into new areas of the temperate regions by the utilization of available cultivars. Subtropical and tropical regions have a great potential for soybean production and that potential will continue to develop as new adapted cultivars are released. Compared to the other major food crops of the world, the improvement of soybeans is still in a juvenile stage.

REFERENCES

Aase, J. K., and Siddoway, F. H. (1976). *Agron. J.* **68**, 627–631.
Abel, G. H., Jr. (1961). *Agron. J.* **53**, 95–98.
Anderson, J. M., and McWhorter, C. G. (1976). *Weed Sci.* **24**, 397–400.
Auclair, J. L. (1959). *Ann. Soc. Entomol. Que.* **5**, 18–43.
Barni, N. A., and Costa, J. A. (1976a). *Agron. Sulriograndense* **12**, 147–162.
Barni, N. A., and Costa, J. A. (1976b). *Agron. Sulriograndense* **12**, 163–172.
Basnet, B., Mader, E. L., and Nickell, C. D. (1974). *Agron. J.* **66**, 531–533.
Boyer, J. S. (1970). *Plant Physiol.* **46**, 233–235.
Bromfield, K. R. (1976). *In* "World Soybean Research" (L. D. Hill, ed.), pp. 491–500. Interstate Printers & Publ., Danville, Illinois.
Brown, D. M. (1960). *Agron. J.* **52**, 493–495.
Bursell, E. (1970). "Introduction to Insect Physiology." Academic Press, New York.

Byth, D. E. (1968). *Aust. J. Agric. Res.* **19,** 879–890.
Camery, M. P., and Weber, C. R. (1953). *Iowa Agric. Exp. Stn., Res. Bull.* **400,** 461–504.
Cartter, J. L., and Hopper, T. H. (1942). *U.S. Dep. Agric., Tech. Bull.* **787,** 66 pp.
Caviness, C. E., and Miner, F. D. (1963). *Agron. J.* **54,** 300–302.
Chamberlain, D. W. (1961). *Phytopathology* **51,** 863–865.
Cooper, R. L. (1971). *Agron. J.* **63,** 449–454.
Dadson, R. B., and Boakye-Boateng, K. B. (1975). *In* "Soybean Production, Protection, and Utilization" (D. K. Whigham, ed.), pp. 69–76. Univ. of Illinois, Urbana-Champaign (INTSOY Ser. 6).
Dart, P., Day, P., Islam, R., and Döbereiner, J. (1975). *In* "Symbiotic Nitrogen Fixation in Plants" (P. S. Nutman, ed.), Int. Biol. Prog., Vol. 7, pp. 361–384. Cambridge Univ. Press, London and New York.
Daugherty, D. M., Neustadt, M. H., Gehrke, C. W., Cavanah, L. E., Williams, L. F., and Green, D. E. (1964). *J. Econ. Entomol.* **57,** 719–722.
Dhingra, O. D., Nicholson, J. F., and Sinclair, J. B. (1973). *Plant Dis. Rep.* **57,** 185–187.
Dorworth, C. E., and Christensen, C. M. (1968). *Phytopathology* **58,** 1457–1459.
Doss, B. D., Pearson, R. W., and Rogers, H. T. (1974). *Agron. J.* **66,** 297–299.
Dusek, D. A., Musick, J. T., and Porter, K. B. (1971). *Tex. Agric. Exp. Stn. MP* **973,** 1–9.
Edwards, C. J., and Hartwig, E. E. (1971). *Agron. J.* **63,** 429–430.
Fairbourn, M. L. (1973). *Agron. J.* **65,** 925–928.
Fausey, N. R., and Schwab, G. O. (1969). *Agron. J.* **61,** 554–557.
Frank, A. B., Harris, D. G., and Willis, W. O. (1974). *Crop Sci.* **14,** 761–765.
Fritschen, L. J. (1966). *Agron. J.* **58,** 339–342.
Garner, W. W., and Allard, H. A. (1920). *J. Agric. Res.* **18,** 553–606.
Garner, W. W., and Allard, H. A. (1923). *J. Agric. Res.* **23,** 871–920.
Gates, D. M. (1965). *Ecology* **46,** 1–13.
Ghorashy, S. R., Monroe, R. L., and Pendleton, J. W. (1971). *Ill. Agric. Exp. Stn., Ill. Res.* **13,** 5–16.
Gilman, D. F., Fehr, W. R., and Burris, J. S. (1973). *Crop Sci.* **13,** 246–249.
Grable, A. B., and Danielson, R. E. (1965). *Soil Sci. Soc. Am., Proc.* **29,** 12–18.
Green, D. E., and Pinnell, E. L. (1968). *Crop Sci.* **8,** 11–15.
Gupta, P. C., and Hittle, C. N. (1970). *Agron. Abstr.* p. 50.
Haines, E. (1955). *Nature (London)* **175,** 474–475.
Hammerton, J. L. (1972). *Exp. Agric.* **8,** 333–338.
Hanks, R. J., and Thorp, F. C. (1957). *Soil Sci. Soc. Am., Proc.* **21,** 357–359.
Hartwig, E. E. (1970). *Trop. Sci.* **12,** 47–53.
Henderson, D. W., and Miller, R. J. (1973). *Calif. Agric. Exp. Stn., Bull.* **862,** 34–40.
Hobbs, P. R., and Obendorf, R. L. (1972). *Crop Sci.* **12,** 664–667.
Hofstra, G., and Hesketh, J. D. (1969). *Planta* **85,** 228–237.
Huang, C., Boyer, J. S., and Vanderhoef, L. N. (1975). *Plant Physiol.* **56,** 222–227.
Hunter, J. R., and Erickson, A. E. (1952). *Agron. J.* **44,** 107–109.
Johnston, T. J., Pendleton, J. W., Peters, D. B., and Hicks, D. R. (1969). *Crop Sci.* **9,** 577–581.
Joyce, R. J. V. (1976). *In* "Insect Flight" (R. C. Rainey, ed.), pp. 135–155. Blackwell, Oxford.
Kalton, R. R., Weber, C. R., and Eldredge, J. C. (1949). *Iowa Agric. Exp. Stn., Res. Bull.* **359,** 733–796.
Knake, E. L., and Slife, F. W. (1962). *Weeds* **10,** 26–29.
Kramer, P. J. (1951). *Plant Physiol.* **26,** 722–736.
Krober, O. A. (1956). *J. Agric. Food Chem.* **4,** 254–257.
Lamb, J., and Chapman, J. E. (1943). *Agron. J.* **35,** 567–578.
Lewis, T. (1964). *Ann. Appl. Biol.* **53,** 165–170.

Matson, A. L. (1964). *Agron. J.* **56**, 552-555.
Metha, A. P., and Prihar, S. S. (1973). *Indian J. Agric. Sci.* **43**, 45-49.
Meyer, R. F., and Boyer, J. S. (1972). *Planta* **108**, 77-87.
Miner, F. D. (1966). *Arkansas, Agric. Exp. Stn., Bull.* **708**, 1-40.
Minor, H. C. (1976). *In* "Expanding the Use of Soybeans" (R. M. Goodman, ed.), pp. 56-62. Univ. of Illinois, Urbana-Champaign (INTSOY Ser. 10).
Nave, W. R., and Wax, L. M. (1971). *Weed Sci.* **19**, 533-535.
Obendorf, R. L., and Hobbs, P. R. (1970). *Crop Sci.* **10**, 563-566.
Ohmura, T., and Howell, R. W. (1960). *Plant Physiol.* **35**, 184-188.
Pendleton, J. W. and Hartwig, E. E. (1973). *In* "Soybeans: Improvement, Production, and Uses" (B. E. Caldwell *et al.*, eds.), pp. 221-237. Am. Soc. Agron., Madison, Wisconsin.
Prine, G. M., West, S. H., and Hinson, K. (1964). *Agron. J.* **56**, 594-595.
Radke, J. K., and Burrows, W. C. (1970). *Agron. J.* **62**, 424-429.
Radke, J. K., and Hagstrom, R. T. (1973). *Crop Sci.* **13**, 543-548.
Sinclair, J. B., and Shurtleff, M. C., eds. (1975). "Compendium of Soybean Diseases." Am. Phytopathol. Soc., St. Paul, Minnesota.
Sorrells, M. E., and Pappelis, A. J. (1976). *Crop Sci.* **16**, 413-415.
Spooner, A. E. (1961). *Arkansas Agric. Exp. Stn., Bull.* **644**, 1-27.
Staniforth, D. W. (1962). *Agron. J.* **54**, 11-13.
Thomas, G. D., Ignoffo, C. M., Biever, K. D., and Smith, D. B. (1974a). *J. Econ. Entomol.* **67**, 683-685.
Thomas, G. D., Ignoffo, C. M., Morgan, C. E., and Dickerson, W. A. (1974b). *J. Econ. Entomol.* **67**, 501-503.
Todd, J. W., and Morgan, L. W. (1972). *J. Econ. Entomol.* **65**, 567-570.
Todd, J. W., and Turnipseed, S. G. (1974). *J. Econ. Entomol.* **67**, 421-426.
Tsiang, T. C. (1948). *In* "Soil Conservation, an International Study," pp. 83-84. FAO, United Nations, Washington, D.C.
Tugwell, P., and Miner, F. D. (1967). *Arkansas Farm Res.* **16**, 12.
Turnipseed, S. G. (1972). *J. Econ. Entomol.* **65**, 224-229.
Turnipseed, S. G. (1973. *In* "Soybean: Improvement, Production and Uses" (B. E. Caldwell *et al.*, eds.), pp. 545-572. Am. Soc. Agron., Madison, Wisconsin.
Turnipseed, S. G., and Kogan, M. (1976). *Annu. Rev. Entomol.* **21**, 247-282.
Whigham, D. K. (1976a). *In* "Expanding the Use of Soybeans" (R. M. Goodman, ed.), pp. 34-37. Univ. of Illinois, Urbana-Champaign (INTSOY Ser. 10).
Whigham, D. K. (1976b). "International Soybean Variety Experiment." Univ. of Illinois, Urbana-Champaign (INTSOY Ser. 11).
Whigham, D. K., Minor, H. C., and Carmer, S. G. (1978). *Agron. J.* (in press).
Whitt, D. M. (1954). *Soybean Dig.* **19**, 10-11.
Willatt, S. T. (1966). *Rhod., Zambia, Malawi J. Agric. Res.* **4**, 95-105.
Woodward, R. G., and Begg, J. E. (1976). *Aust. J. Agric. Res.* **27**, 501-508.

5

Breeding

WALTER R. FEHR

I.	Breeding Objectives	120
	A. Seed for Commercial Production	120
	B. Characters, Inheritance, and Source of Genes	121
II.	Steps in Cultivar Development	127
	A. Selection of Parents	128
	B. Development of Pure Lines	128
	C. Testing of Lines	129
	D. Purification and Release	130
III.	Breeding Methods	132
	A. Backcrossing	132
	B. Pedigree Selection	135
	C. Single-Seed Descent	136
	D. Early-Generation Testing	138
	E. Recurrent Selection	139
	F. Comparison of Alternative Breeding Methods	141
IV.	Breeding Operations	143
	A. Crossing Procedures	143
	B. Generation Advance	146
	C. Yield Testing	146
	D. Plot Size and Shape	147
	E. Field Operations and Equipment	150
V.	Blends	152
VI.	Hybrids	153
	References	155

I. BREEDING OBJECTIVES

A. Seed for Commercial Production

Soybean breeding has played a significant role in the improvement of soybean production. Cultivars grown by farmers today have been improved for yield potential, shattering resistance, disease resistance, and other characters.

The soybean cultivars grown by farmers in the United States up to the 1940's were introduced from Asia (Hartwig, 1973). Introductions were tested for agronomic performance, and the superior ones were released to farmers. When an introduction was a mixture of different genetic types, individual plants were selected and tested, and the superior ones were made available for commercial production.

The best introductions were used as parents to develop superior cultivars and are the ancestors of our current cultivars. For example, BEESON is a current cultivar that has two introductions, RICHLAND and MUKDEN, as grandparents. RICHLAND and MUKDEN were grown as commercial cultivars before new cultivars were developed by modern soybean breeding methods.

Soybean cultivars are considered true-breeding because the seed harvested is the same genetically as the seed planted. The cultivars generally are grown individually in pure stand. However, in recent years, seedsmen and farmers have mixed seed of cultivars to form blends. A description of soybean blends is presented in Section V.

Hybrid soybeans are not yet commercially available, however, a United States patent has been issued for a system of hybrid seed production that the inventors believe may be successful in the future. The possibilities and implications of hybrid seed production are described in Section VI.

There has been an increase in the number of new cultivars available for commercial production as a result of increased effort by both public and private agencies. The change in cultivars can be illustrated by comparing cultivars grown in 1968 with those in 1976 as reported by the United States Department of Agriculture Statistical Reporting Service from data collected in fourteen northern and southern states. Cultivars grown on more than 2% of the acreage in 1968 were WAYNE (14.3%), AMSOY (13.0%), LEE (11.1%), CLARK and CLARK 63 (10.5%), HAROSOY and HAROSOY 63 (8.7%), CHIPPEWA and CHIPPEWA 64 (7.7%), BRAGG (6.5%), HILL (4.0%), HAWKEYE and HAWKEYE 63 (3.3%), HAMPTON (3.0%), and HARK (2.3%). Cultivars grown on more than 2% of the acreage in 1976 were WILLIAMS (8.5%), BRAGG (7.9%), FORREST (7.3%), CORSOY (7.1%), WAYNE (5.1%), LEE (4.9%), AMSOY and AMSOY 71 (4.7%), PICKETT and PICKETT 71 (4.0%), DAVIS (3.9%), DARE (2.9%), CUTLER and CUTLER 71 (2.4%), CALLAND (2.3%), WELLS (2.1%),

5. Breeding

BEESON (2.0%), CLARK and CLARK 63 (2.0%), and RANSOM (2.0%). There are only five cultivars that appear on both the 1968 and 1976 lists, WAYNE, AMSOY, LEE, CLARK, and CLARK 63. The most widely grown cultivar in 1976, WILLIAMS, was not even available for planting in 1968.

B. Characters, Inheritance, and Source of Genes

The adaptation of soybean cultivars is strongly influenced by their photoperiodic response (Chapter 2, Section II, A; Chapter 3, Section II) and as a result, cultivars grown in the United States generally are most productive in a relatively narrow range of latitudes. Each breeder must decide which characters need greatest attention in the geographical area where the new cultivar will be grown. A breeder in the northern United States may consider increased seed yield and phytophthora rot resistance as the two most important characters. A breeder in the southern United States may have to consider additional factors such as shattering, seed quality, and nematodes.

Cultivars frequently have many characters that could be improved, but only a few can be chosen for major attention. The more characters a breeder must consider in the development of a new cultivar, the lower will be the chance of making maximum progress for any single trait. For example, assume we want to develop a cultivar with high protein, lodging resistance, resistance to phytophthora rot, and high yield. Assume that the frequency of lines with the desired protein level is 1 out of 10, lodging resistance 1 of 7, resistance to phytophthora rot 1 of 4, and high yield 1 of 50. If the characters are inherited independently, an average of only 1 out of every 14,000 individuals would have all the desired characteristics. If the breeder limited the objective to improvement of only protein level and lodging resistance, the frequency of desired individuals would increase to 1 out of 70.

The selection of an appropriate breeding procedure depends on the inheritance of the character to be improved. Characters are placed in two broad categories based on their inheritance, qualitative or quantitative (Allard, 1960). Qualitative characters are those that have distinctly different types that can be readily distinguished. They also are referred to as simple traits because they are controlled by one or a few genes. Pubescence color is an example of a qualitative character. It has two distinctly different types, brown and gray, and is controlled by one gene. Qualitative characters common in commercial cultivars are listed in Tables I and II.

Quantitative characters have variation in expression that cannot be separated into distinct classes. They are called complex characters because they are controlled by many genes. The variation within a quantitative character is due to its complex inheritance and to the influence of the environment. Plant height is a quantitative character, and soybean lines range in height

TABLE I

Inheritance of Some Qualitative Characters Common in Soybean Cultivars[a]

Character	Gene	Cultivar or strain
Flower color		
Purple	W_1	AMSOY 71
White	w_1	WAYNE
Pubescence color		
Tawny (brown)	T	WAYNE
Gray	t	AMSOY 71
Pod color		
Black	$L_1 L_2$	SENECA
Black	$L_1 l_2$	T215[c]
Brown	$l_1 L_2$	CLARK
Tan	$l_1 l_2$	DUNFIELD
Seed coat color		
Yellow	I or i^1	AMSOY
Dark-colored	i	SOYSOTA[b]
Stem termination		
Indeterminate	Dt_1	CLARK
Determinate	dt_1	EBONY
Semi-determinate	Dt_2	T117
Indeterminate	dt_2	CLARK
Time of flowering and maturity		
Late	E_1	T175
Early	e_1	CLARK
Late	E_2	CLARK
Early	e_2	T245
Late and sensitive to fluorescent light	E_3	HAROSOY 63
Early and insensitive	e_3	BLACKHAWK
Seed coat luster		
Dull	$B_1 B_2 B_3$	SOOTY
Shiny	b_1, b_2 or b_3	WAYNE
Leaflet form		
Broad leaflet	Ln	AMSOY 71
Narrow leaflet	ln	T41

[a] Adapted from Bernard and Weiss, 1973.

[b] Seeds with black or brown seed coats occasionally occur in commercial cultivars. It is due to a mutation of the genes I or i^1 to i (Bernard and Weiss, 1973).

[c] Strains designated with a T are available from the Soybean Genetic Collection maintained at Urbana, Illinois.

from less than 25 cm to more than 125 cm. A cultivar may be 90 cm tall in a good environment, but only 60 cm tall under poor growing conditions. Common quantitative characters include seed yield, lodging resistance, protein and oil percentage, and seed size.

Sources of genes for qualitative and quantitative characters include commercial cultivars, experimental lines, and plant introductions. Over 5000 introductions of soybeans from foreign countries are maintained by the United States Department of Agriculture at Urbana, Illinois and Stoneville, Mississippi. They include lines collected by United States scientists in foreign countries and lines sent to the United States by foreign scientists. Qualitative genes also are available in lines maintained in the Genetic Type Collection of the United States Department of Agriculture at Urbana, Illinois (Bernard and Weiss, 1973). The plant introductions and genetic collection are valuable to breeders, geneticists, physiologists, and other research scientists in public and private institutions.

Reviews of important soybean characters were made by Bernard and Weiss (1973), Brim (1973), and Hartwig (1973). General comments about a few of the characters will illustrate the range of possibilities for future improvement in cultivars.

1. Seed Yield

Seed yield is the most important character in soybean breeding. It is a quantitative character that is strongly influenced by the envirinment (Chapter 4). The most widely used source of parents for yield improvement is high-yielding cultivars and experimental lines; however, plant introductions are being investigated as a source of new genes for yield improvement.

2. Lodging Resistance

Excessive lodging can be detrimental to seed yield and can make harvest difficult. Two approaches are being used to improve lodging resistance. The most common procedure is to select for lodging resistance among soybean lines with height similar to current cultivars. A second procedure being studied in the northern United States is to control lodging by reducing plant height. Both semi-determinate and determinate stem termination are being evaluated for reduction of plant height. The semi-determinate character is controlled by a dominant gene, Dt_2, and the determinate character by a partially recessive gene, dt_1 (Table I). The Dt_2 gene is available from the Soybean Genetic Collection and the dt_1 gene is present in most cultivars adapted to the southern United States.

TABLE II

Inheritance of Hilum Color in Seeds with a Yellow Seed Coat[a]

	Hilum Color	Pubescence color	Flower color
Gray	Tawny	Purple
Gray	Tawny	White
Gray	Gray	Purple
Yellow	Gray	White
Gray	Tawny	Purple
Gray	Tawny	White
Gray	Gray	Purple
Yellow	Gray	White
Yellow	Tawny	Purple
Yellow	Tawny	White
Yellow	Gray	Purple
Yellow	Gray	White
Yellow	Tawny	Purple
Yellow	Tawny	White
Yellow	Gray	Purple
Yellow	Gray	White

[a] The genes I, R, and O control the distribution and color of pigmentation in the seed. The genes for pubescence (T,t) and flower color (W_1, w_1) also are part of the genetic system

3. Protein and Oil Content

The protein percentage in soybean seed is about 40% and the oil percentage about 21% in most commercial cultivars. Increasing protein or oil production per unit land area can be accomplished in three ways; (1) increase seed yield while maintaining a constant protein or oil percentage, (2) increase protein or oil percentage while maintaining a constant seed yield, or

5. Breeding

TABLE II *(Continued)*

Inheritance of Hilum Color in Seeds with a Yellow Seed Coat[a]

Hilum Color	Pubescence color	Flower color
Black	Tawny	Purple
Black	Tawny	White
Imperfect black	Gray	Purple
Buff	Gray	White
Black	Tawny	Purple
Black	Tawny	White
Imperfect black	Gray	Purple
Buff	Gray	White
Brown	Tawny	Purple
Brown	Tawny	White
Buff	Gray	Purple
Buff	Gray	White
Reddish-brown	Tawny	Purple
Reddish-brown	Tawny	White
Buff	Gray	Purple
Buff	Gray	White

controlling hilum color. To determine the genes controlling a particular hilum color, begin with W_1 or w_1 and follow the line back to I or i^i. Several different gene combinations can result in the same hilum color. (Adapted from R. G. Palmer, personal communication.)

(3) increase both seed yield and protein or oil percentage. Soybeans currently are marketed by weight without regard to protein or oil content; therefore, farmers do not consider seed composition in the selection of a cultivar for production. Consequently, soybean breeders generally develop cultivars with high yield without major attention to protein or oil content.

Breeding for both higher protein and higher oil content simultaneously has not been successful because the characters are negatively associated. An increase in protein percentage generally is associated with a decrease in the oil percentage (Brim, 1973). Plant introductions are the primary source of genes for unusually high protein or oil percentage.

4. Protein and Oil Quality

The primary interest in protein quality is to increase the percentage of the amino acid methionine. There is little breeding effort on protein quality because there are no cultivars or plant introductions available with a high level of methionine. The high cost of analysis for methionine content also is a deterrent to breeding efforts.

Oil quality may be improved by decreasing one of the components in the oil, linolenic acid, to less than 3%. Present commercial cultivars have about 7% linolenic acid, plant introductions have about 5%, and breeding programs at Iowa State University and North Carolina State University have developed lines with about 4%. The market value for seed with improved oil quality has not been determined.

5. Seed Size

Most United States cultivars have a seed size of 12–19 g per 100 seeds, but there is a limited demand for cultivars with a seed size greater than 20 g per 100 seeds. The large seeds are preferred for certain foods in the Orient and are used to a limited extent for direct human consumption in the United States. Large-seeded cultivars are grown by farmers who have a contract with a company for purchase of the seed. They are not preferred for widespread commercial production because seedlings from large seeds have difficulty emerging from the soil and seeds are more susceptible to cracking during harvest.

Genes for large seed size are available from current large-seeded cultivars and plant introductions. A plant introduction with up to 55 g per 100 seeds is available (Hartwig, 1973).

6. Resistance to Phytophthora Rot

Phytophthora rot (*Phytophthora megasperma* Drechs. var. *sojae* Hild.) is one of the major soybean diseases in the United States. Until the early 1970's only race 1 of the fungus was considered to be important. Race 2 of the fungus had been identified, but was not of economic importance. Since the early 1970's, new races of economic importance have been identified in Ohio, Indiana, and surrounding areas (Hartwig, 1973).

The most common strategy used to control the disease has been to use genes for specific resistance, sometimes called vertical resistance. The

5. Breeding

strategy is popular because resistant plants are not injured by the organism, the genes can be transferred readily from one cultivar to another by backcrossing, and laboratory techniques are available that clearly distinguish resistant and susceptible plants. Specific resistance to race 1 is controlled by a single dominant gene (Bernard and Weiss, 1973) and has been used to develop resistant cultivars such as AMSOY 71 and PICKETT 71. Specific resistance to some of the new races has been found in cultivars, such as MACK and TRACY, and in some plant introductions. Inheritance of resistance to the new races is being investigated with preliminary results from backcrossing programs, suggesting that resistance may be controlled by a few major genes.

A second strategy for minimizing yield loss from phytophthora rot is to develop cultivars with general resistance to the fungus. General resistance sometimes is referred to as field resistance, horizontal resistance, or tolerance. Cultivars with general resistance are not immune to the organism and can be killed by artificial inoculation in the laboratory. Under field conditions, however, the cultivars may be infected by the fungus, but do not suffer as serious a yield reduction as more susceptible cultivars.

The advantage of cultivars with general resistance is that they may be able to withstand new races of the disease organism, thereby increasing the length of their usefulness in commercial production. Cultivars with specific resistance may have a short life span if new races of the fungus become important to which they are susceptible. For example, cultivars with specific resistance to race 1 are no longer useful in fields where race 3 is prevalent. If cultivars with specific resistance to race 3 are planted, a new race may increase in importance.

Breeding for cultivars with general resistance has been limited because of difficulty in evaluation of the character. Evaluation for general resistance currently must be done in the field, but such tests are difficult because severity of the disease is not uniform throughout a field. Expression of the disease is influenced by weather conditions; therefore, a location may not be useful for testing every year.

Inheritance of general resistance has not been determined, but it seems to be a quantitative character. There has not been adequate testing to determine all the useful sources of genes, but cultivars such as WAYNE and FORREST are considered to have general resistance and may be useful as parents.

II. STEPS IN CULTIVAR DEVELOPMENT

Cultivar development consists of four steps: selection and crossing of parents, development of pure lines, testing of the lines, and purification and release of a new cultivar.

A. Selection of Parents

Selection of appropriate parents will ultimately determine the success of a breeding program. Whenever possible the breeder uses at least one parent that has an adequate level of a character desired in the progeny, particularly for qualitative characters. For example, resistance to race 1 of *Phytophthora* is controlled by a single gene. If the breeder wants to select progeny with resistance, at least one of the parents must have the desired gene.

Selection of appropriate parents for quantitative characters is difficult because the breeder does not know what genes are present in a potential parent. Nevertheless, the chances for success are greatest when parents are chosen that are the best available for the character being considered. For example, a breeder may want to improve lodging resistance. Parents available include cultivars and plant introductions that range from very susceptible to moderately resistant. There is a possibility that the cross of a lodging susceptible parent with a moderately resistant one will provide offspring with superior lodging resistance. The odds of success are better, however, if two cultivars with the best lodging resistance available are crossed.

B. Development of Pure Lines

The cultivars currently grown by farmers are true-breeding pure lines. Breeders use the term "pure line" to describe both tested and untested true-breeding lines, and the terms "variety" or "cultivar" (cultivated variety) to describe pure lines used in commercial production. Every plant in a pure line has a similar genetic makeup because it is developed and maintained by self-pollination. Self-pollination occurs when the pollen (male) from a flower fertilizes an ovule (female) on the same plant. This process occurs naturally in soybeans because the male and female organs are in the same flower.

The development of a cultivar begins with the crossing of parents to obtain hybrid F_1 seed (Section IV, A). The F_1 plants are heterozygous for all genes that were different in the parents. For example, if the parents differed in pubescence color, the hybrid plants would be heterozygous because two forms (alleles) of the gene are present (Fig. 1). The F_1 plant would have brown pubescence because the allele for brown (T) is dominant to the allele for gray (t).

Each F_2 (second generation) seed harvested from an F_1 plant probably has a different genetic makeup. If two parents are crossed that differ by only 15 genes, there are 14,348,907 different F_2 seeds that could be produced. The number of different F_2 individuals available to the breeder from most crosses, therefore, is extremely large because most parents differ by many more than 15 genes.

5. Breeding

Generation	Percentage of individuals		
	TT	Tt	tt
F_1	0	100	0
F_2	25	50	25
F_3	37.5	25	37.5
F_4	43.75	12.5	43.75
F_5	46.875	6.25	46.875

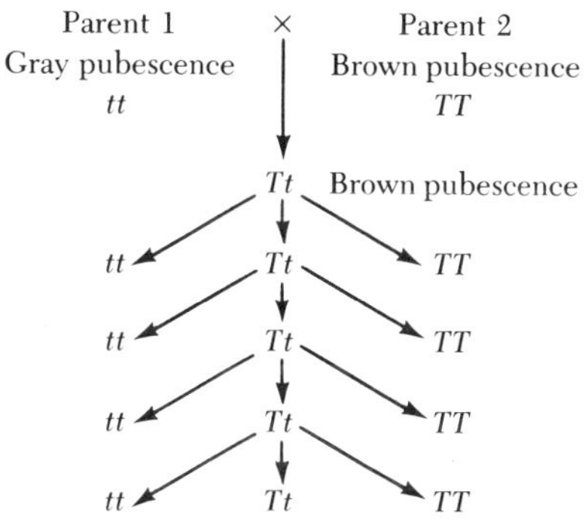

Parent 1 Gray pubescence *tt* × Parent 2 Brown pubescence *TT*
↓
Tt Brown pubescence
↓
tt ← *Tt* → *TT*
↓
tt ← *Tt* → *TT*
↓
tt ← *Tt* → *TT*
↓
tt ← *Tt* → *TT*

Fig. 1. Segregation of a single gene from a two-way cross. The percentage of true-breeding lines (*tt* and *TT*) increases with each generation of self-pollination.

There are few, if any, F_2 plants that are true-breeding for all genes; however, many would be true-breeding for at least one gene. Self-pollination of an F_1 plant would result in the *T* or *t* allele from the male pairing with the *T* or *t* of the female to produce 50% heterozygous (*Tt*) F_2 seeds, 25% homozygous *TT*, and 25% homozygous *tt* (Fig. 1). When the F_2 plants are grown, 75% on the average would have brown pubescence (50% *Tt* and 25% *TT*) and 25% gray pubescence. The F_3 seed produced on heterozygous F_2 plants would be like that produced by the F_1 hybrid (50% *Tt*, 25% *TT*, and 25% *tt*), however, the *TT* F_2 plants could produce only *TT* seeds and the *tt* plants only *tt* seeds. Therefore, with each generation of self-pollination the percentage of heterozygous individuals decreases and the percentage of true-breeding individuals increases.

C. Testing of Lines

Breeding methods differ in the amount of evaluation that takes place during the generations of selfing. With pedigree selection, evaluation generally begins in the F_2 generation (Section III, B); however, evaluation of lines developed by single-seed descent may not begin before the F_5 generation (Section III, C). When selection is carried out during the early selfing generations, the goal is to have an elite group of pure lines available for extensive testing.

Pure lines developed by any method must be evaluated before they can be considered for release to farmers. The amount of testing required depends on the influence of the environment on expression of the character. A qualitative character that is not influenced by the environment only requires limited evaluation. Specific resistance to race 1 of *Phytophthora* is a qualitative character that can be evaluated in a single laboratory test. A quantitative character that is strongly influenced by the environment requires extensive evaluation. Seed yield is a character that is strongly influenced by environment, and pure lines must be tested over many different locations and years to determine their yield potential (Section IV, C).

D. Purification and Release

A line that has been judged suitable for release is purified to remove off-type plants that can arise by several means. (1) The off-types may represent a few seeds from another line that were mixed during threshing. (2) Natural crossing may occur between lines grown adjacent to one another, and the hybrid and its offspring would represent off-types. (3) A line may have been heterozygous for a gene when it was selected for testing. Progeny within the line would segregate for the gene in the same manner as from a F_1 hybrid. For example, lines may be uniform for height, maturity, and other characters, but nonuniform for flower color. One of the two flower colors would have to be removed as an off-type. (4) Natural genetic changes (mutations) can cause visible changes in plant or seed characteristics. A relatively common mutation is from yellow seed coat to brown or black seed coat (Table I). Off-types caused by mutation would be removed during purification.

A reliable procedure for purifying and increasing a line is outlined in Fig. 2. Individual plants with uniform characteristics are harvested from a line, and the seed from each plant is grown in a separate row the following generation. Any row that has off-type plants or is not similar to the other rows is discarded as an off-type. Each row is harvested separately and seed from each row is inspected for uniformity of seed coat color, hilum color, seed coat luster, and also may be tested for disease resistance, protein content, or other characters. Seeds from uniform rows are bulked to obtain pedigree seed that is used for increase of the cultivar. The few kilograms of pedigree seed must be increased to the tons of seed required for commercial production. Each generation of increase requires care in avoiding seed mixtures or natural crossing with other cultivars. Each generation of seed increase is given a separate name. Pedigree seed (30 kg) is used to produce breeder seed (3 tons), followed by foundation seed (180 tons), registered seed (5400 tons), and certified seed (162,000 tons). The numbers in paren-

5. Breeding

Season	Procedure
1	Harvest individual plants
2	Plant individual rows Discard off-type rows Bulk seed of similar ones
3	Plant pedigree seed Rogue off-type plants Harvest breeder seed
4	Plant breeder seed Rogue off-type plants Harvest foundation seed
5	Plant foundation seed Harvest registered seed
6	Plant registered seed Harvest certified seed

Pedigree seed—30 kilograms — 1.5 hectares
Breeder seed—3 tons — 100 hectares
Foundation seed—180 tons — 3,000 hectares
Registered seed—5,400 tons — 90,000 hectares
Certified seed—162,000 tons

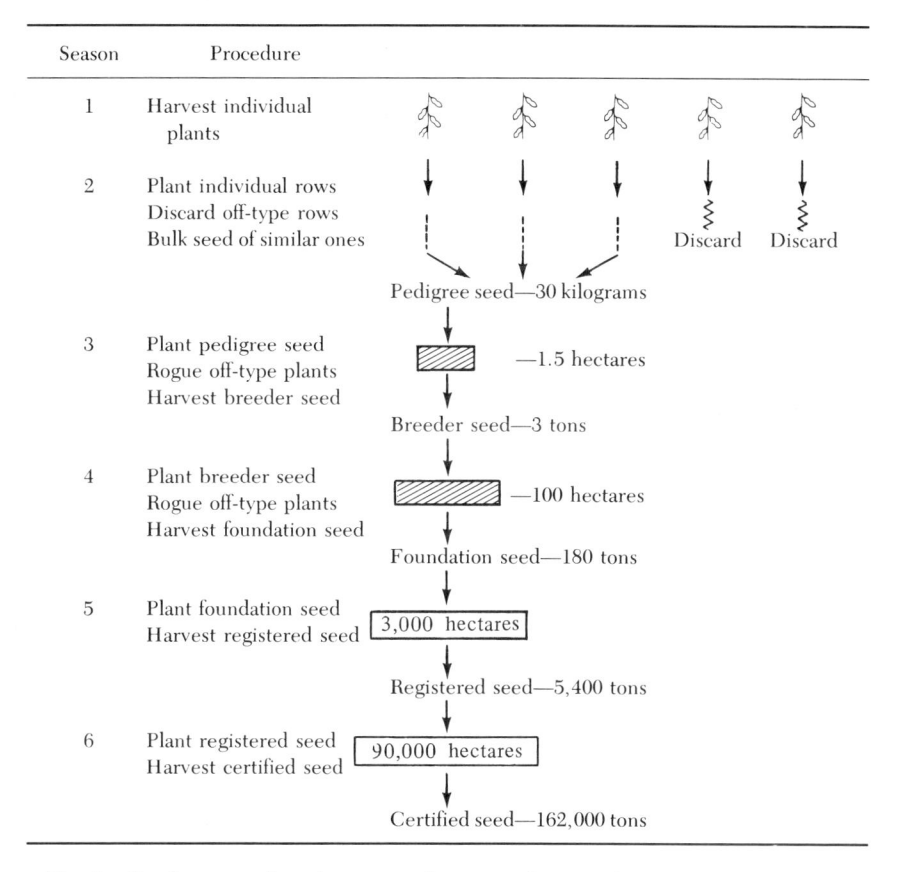

Fig. 2. Purification and seed increase of a new cultivar. Kilograms, tons, and hectares indicated are approximate quantities that could be produced for each seed class.

theses are examples of the approximate quantities of seed that could be produced for each seed class.

Production of pedigree and breeder seed generally is under the supervision of the breeder. Foundation seed production is handled by an organization within a company or institution that specializes in large seed increases. Production of registered and certified classes is by seedsmen who sell the seed to the farmers. Seed certification is supervised and regulated by an official agency in each state. Requirements for certification of soybeans are established by the state in cooperation with the American Association of Official Seed Certifying Agencies.

The purpose of certification is to maintain the genetic identity of a cultivar during seed production and distribution. A farmer that purchases certified seed of the cultivar LEE is assured that the seed has met the standards for genetic purity. The percentage of soybeans planted with certified seed can be estimated by comparing the number of seed units (bushels) certified with the commercial area planted the following year. In Iowa, enough seed was certified in 1959 to plant 8% of the soybean area in 1960, 7% in 1965, 27% in 1970, and 41% in 1975. The large increase in use of certified seed by Iowa farmers in recent years is representative of the trend in other states.

The developer of a new soybean cultivar can obtain exclusive rights to its sale and distribution for a period of 17 years under the United States Plant Variety Protection Act established in 1970. There are two options for protection. (1) The owner who specifies that the cultivar may be sold either as certified or noncertified seed assumes responsibility for prosecuting anyone that illegally sells or distributes seed. (2) The owner may stipulate that seed may be sold by cultivar name only as a class of certified seed and violators are subject to prosecution by the federal government under the Federal Seed Act. Seed certification, therefore, may be used by breeders to protect their right of ownership.

III. BREEDING METHODS

Several methods are used by breeders to develop soybean cultivars. They are modified to fit the resources available; therefore, only a general description of the methods is possible (Allard, 1960).

A. Backcrossing

Backcrossing is used to transfer one or a few genes from one cultivar to another (Fig. 3). It has been used successfully to transfer genes controlling specific resistance to phytophthora rot, cyst nematode, and root-knot nematode to commercial cultivars that were susceptible. A new cultivar developed by backcrossing frequently is designated by the name of the recurrent parent and the year the resistant cultivar was released. For example, CHIPPEWA 64 was developed with CHIPPEWA as the recurrent parent and was released for commercial production in 1964. Other examples of cultivars developed by backcrossing include AMSOY 71, CUTLER 71, and HOOD 75.

The method is initiated by making a cross between two parents to produce a hybrid F_1 (Fig. 3). The F_1 plant is crossed back to the parent that is being improved, called the recurrent parent because it recurs or is repeatedly used for crossing. The parent that contributes the desired gene is the donor parent, or nonrecurrent parent because it is used to make the initial cross,

5. Breeding

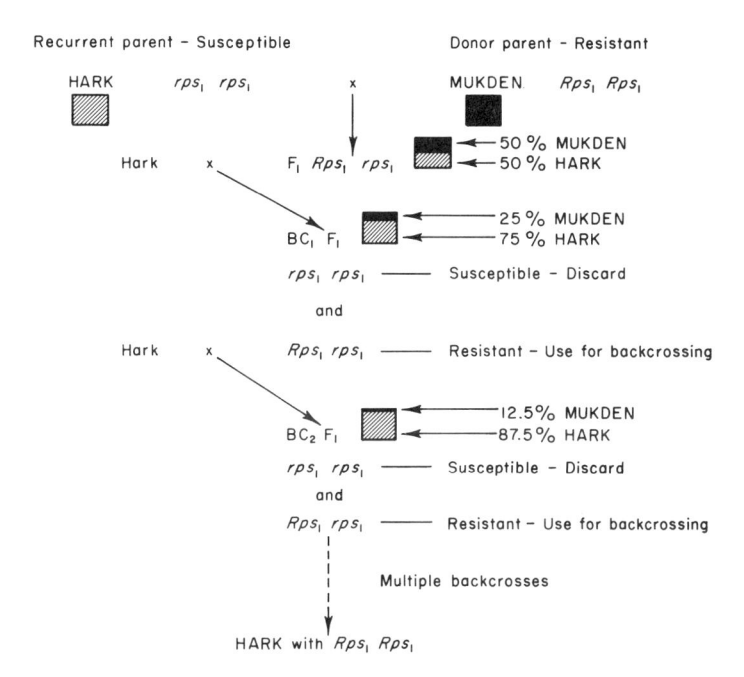

```
Recurrent parent - Susceptible          Donor parent - Resistant
   HARK        rps₁ rps₁         x         MUKDEN       Rps₁ Rps₁
                                                    ← 50% MUKDEN
       Hark    x      F₁ Rps₁ rps₁              ← 50% HARK

                                                 ← 25% MUKDEN
                     BC₁ F₁                      ← 75% HARK
                     rps₁ rps₁  ——— Susceptible - Discard
                         and
       Hark    x        Rps₁ rps₁ ——— Resistant - Use for backcrossing

                                                 ← 12.5% MUKDEN
                     BC₂ F₁                      ← 87.5% HARK
                     rps₁ rps₁  ——— Susceptible - Discard
                         and
                     Rps₁ rps₁  ——— Resistant - Use for backcrossing
                         ⋮
                     Multiple backcrosses
                         ↓
                  HARK with Rps₁ Rps₁
```

Fig. 3. Schematic representation of transferring a gene from one cultivar to another by backcrossing. The donor parent MUKDEN has resistance to race 1 of phytophthora rot (*Phytophthora megasperma* var. *sojae*) to which HARK is susceptible. When backcrossing is completed, HARK should have almost 100% of its original genes and the gene for resistance to phytophthora rot.

but does not recur in the backcrossing program. The purpose of crossing back to the recurrent parent is to recover all of its desirable genes. Each time a cross is made back to the recurrent parent an additional 50% of its genes are recovered. The breeder will continue backcrossing until the desired level of genes from the recurrent parent has been recovered.

As the backcross program progresses, the breeder must be certain that the desired genes from the donor parent are being retained. A single gene controlling a qualitative character can be readily monitored; however, it would be difficult to monitor the many genes controlling a quantitative character. Backcrossing, therefore, is used primarily for transferring characters controlled by few genes.

The primary limitation of the backcross method is that only one character is improved in the new cultivar. The yield, lodging resistance, shattering resistance, and other characteristics of the new cultivar generally will be the same as the original cultivar.

Season	Procedure	Type of population
		Two-way cross (Single cross, Two-parent cross)
1	Cross two parents Obtain hybrid F_1 seed	Parent 1 × Parent 2 ↓
2	Grow F_1 plant Obtain F_2 seed	F_1 │ Self-pollination ↓ F_2 seed
		Three-way cross (Three-parent cross)
1	Cross two parents Obtain hybrid F_1 seed	Parent 1 × Parent 2 ↓
2	Grow F_1 plant Cross F_1 to third parent Obtain hybrid F_1 seed	F_1 plant × Parent 3 ↓ F_1 │ Self-pollination ↓
3	Grow F_1 plants Obtain F_2 seed	F_2 seed
		Intermated population (Random mated population)
1	Make two-way crosses between parents (first intermating) Obtain hybrid F_1 seed	Parents 1 × 2 3 × 4 5 × 6 7 × 8 ↓ ↓ ↓ ↓
2	Cross F_1 plants from different two-way crosses (second intermating) Obtain hybrid seed	F_1 × F_1 F_1 × F_1
3	Cross F_1 plants from different four-way crosses (third intermating) Obtain intermated seed	F_1 × F_1 ↓ Intermated seed ↓
4	Begin selfing	Plants of intermated population

B. Pedigree Selection

Pedigree selection has been used successfully to improve seed yield, lodging resistance, and many other characters. Selection of desired types generally begins in the F_2 generation and continues until pure lines have been developed. A record (pedigree) is maintained for each line during each generation of selection.

Pedigree selection can be used with any type of population (Fig. 4). Desirable F_2 plants are harvested individually and the F_3 seed from each plant is sown in a separate row (Fig. 5). The breeder compares the rows, selects the ones that are most desirable, and harvests several desirable F_3 plants from each selected row. If the original F_2 plant had few heterozygous genes, the F_3 plants within that family (row) will be similar in their genetic makeup. However, if the F_2 plant had many heterozygous genes, the F_3 plants will be quite different.

A separate row is planted from each F_3 plant the next season. Seed from F_3 plants that trace back to the same F_2 plant (were selected from the same F_2 progeny row) are sown in adjacent rows and are called a family. The breeder first chooses the best families, then the most desirable rows within the best families, and harvests two to four desirable F_4 plants within the selected rows. Seed from each F_4 plant is sown in a separate row the next season adjacent to other rows from the same family (same F_3 row). The breeder selects the best families, the best rows within the best families, and several F_5 plants within the selected rows. The next season a separate row is planted from each F_5 plant adjacent to other rows from the same family. The breeder selects the best families and the best row within the selected families. Many rows from F_5 plants look uniform and selection of individual plants within a line would be of little value. The breeder, therefore, harvests the entire row in bulk and the pure line is ready for extensive evaluation to determine its value as a new cultivar.

Pedigree selection is effective for the improvement of seed yield, lodging resistance, and other characters when the breeder is able to differentiate between superior and inferior plants and progeny rows. Most studies have shown that visual selection for seed yield is effective for eliminating poor lines, but is not effective for differentiating between moderate- and high-yielding lines. Pedigree selection is effective for lodging resistance or other characters that can be readily evaluated by visual examination.

The use of pedigree selection has decreased in recent years as greenhouses and semi-tropical or tropical locations have become widely used

---

Fig. 4. The development of three types of populations used for selection of improved cultivars. The titles in parentheses are sometimes used to describe the populations.

Season	Procedure
1	Plant F_2 seed Select individual plants
2	Grow individual rows Select best rows Select best F_3 plants
3	Grow individual rows Select best families Select best rows Select best F_4 plants
4	Grow individual rows Select best families Select best rows Select best F_5 plants
5	Grow individual rows Select best families Harvest best rows in bulk
6	Extensive testing of F_5-derived lines

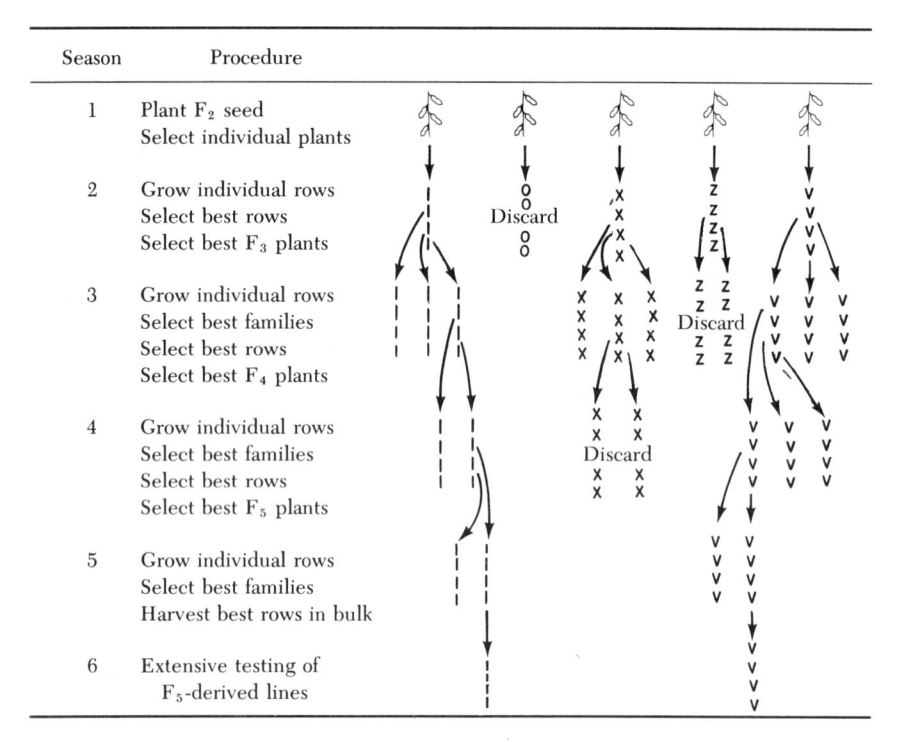

Fig. 5. Illustration of the pedigree selection method.

for growing several generations each year. In winter facilities, differences among lines for yield, lodging resistance, height, and other characters are not readily expressed, and pedigree selection for those characters is not effective.

C. Single-Seed Descent

The most widely used procedure for yield improvement is single-seed descent. Its popularity is related to the use of winter facilities in soybean breeding.

The F_2 seeds from a cross are planted (Fig. 6) and at harvest one random F_3 seed is taken from each F_2 plant. The seeds are bulked together for planting and one random F_4 seed is harvested from each F_3 plant. The process is repeated until the plants are considered sufficiently true-breeding, generally the F_4 or F_5 generation. For the example shown in Fig. 6, F_5 plants were considered sufficiently true-breeding, so individual plants were harvested. The seed from each plant was sown in a separate row, and the desirable rows were selected as lines for extensive testing.

5. Breeding

Season	Procedure					
1	Grow F_2 plants Harvest 1 seed/plant			F_2 plants ↓		
2	Grow F_3 plants Harvest 1 seed/plant			F_3 plants ↓		
3	Grow F_4 plants Harvest 1 seed/plant			F_4 plants ↓		
4	Grow F_5 plants Harvest individual plants	⚘ ↓	⚘ ↓	⚘ ↓	⚘ ↓	⚘ ↓
5	Grow individual rows Harvest selected rows in bulk	| | Discard | |	o o o o ↓	x x Discard x x	z z z z ↓	v v Discard v v
6	Extensive testing of F_5-derived lines		o o o o		z z z z	

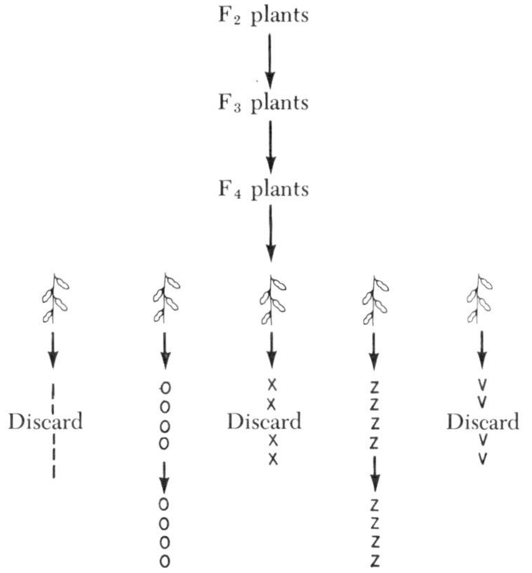

Fig. 6. Illustration of one procedure used for the single-seed descent method.

Single-seed descent is not influenced by the atypical growth that occurs in winter facilities. Only a few seeds are needed from each plant; therefore, total yield or appearance of the plants is of no consequence. Many plants can be crowded into a small space because each must produce only a few seeds for harvest.

Many breeders prefer to harvest several seeds from the plants in each generation, instead of only one. This provides adequate seed for planting and a reserve in case of crop failure. If only one seed per plant is harvested, as in Fig. 6, a separate harvest of a reserve sample must be made.

There is another procedure for obtaining pure lines that can be considered a form of single-seed descent (Fig. 7). Individual F_2 plants are harvested separately and F_3 seeds from each F_2 plant are sown in a small hill or row. F_4 seeds are harvested from each hill, replanted, and F_5 seeds are harvested. The process is repeated for as many generations as the breeder desires. Breeders who use the procedure are assured that each original F_2 plant will be represented by progeny in advanced generations. The procedure takes more time and space than other single-seed descent techniques; therefore, it is used primarily in research studies.

Season	Procedure
1	Grow F$_2$ plants Harvest F$_3$ seed from each plant individually
2	Grow individual hills Harvest F$_4$ seed from each hill individually
3	Grow individual hills Harvest F$_5$ seed from each hill individually
4	Grow individual hills Harvest one F$_5$ plant from each hill
5	Grow individual rows Harvest selected rows in bulk
6	Extensive testing of F$_5$-derived lines

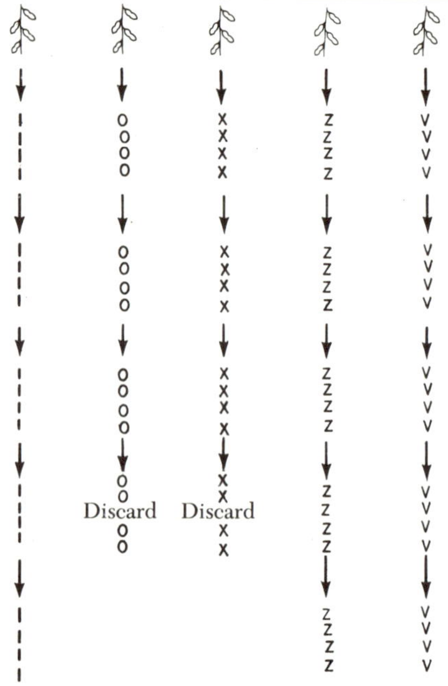

Fig. 7. Illustration of a single-seed descent procedure to assure that each F$_2$ plant is represented by progeny in advanced generations.

D. Early-Generation Testing

Early-generation testing was first suggested as a method of identifying crosses that would contain superior pure-line progeny. Crosses were made between parents to obtain F$_1$ seed, the F$_1$ plants were grown, and the F$_2$ seed from them was used to plant a yield test. Each entry in the yield test was a different cross, and crosses with the highest yield were used as a source of pure-line progeny. The procedure was not widely adopted because research indicated that the number of superior progeny often was as high in low-yielding crosses as in high-yielding ones.

The term early-generation testing generally is associated today with the testing of progeny from F$_2$ plants. Individual F$_2$ plants from a cross are harvested, the F$_3$ seeds from each F$_2$ plant are sown in a separate row, and inferior rows discarded (Fig. 8). The F$_4$ seed from selected rows is used for yield tests of the F$_2$ lines. F$_5$ seeds from the superior F$_2$ lines in the yield test are planted and individual F$_5$ plants harvested. A separate row is grown from each F$_5$ plant and desirable rows are selected as pure lines for extensive testing.

5. Breeding

Season	Procedure
1	Grow F_2 plants Harvest each F_2 plant individually (F_2-derived lines)
2	Grow individual rows Select best rows Harvest F_4 seed of selected rows in bulk
3	Grow replicated yield test with F_4 seed Select highest yielding F_2-derived lines
4	Grow F_5 plants from selected F_2-derived lines Harvest selected F_5 plants individually
5	Grow individual rows Harvest selected rows in bulk
6	Extensive testing of F_5-derived lines

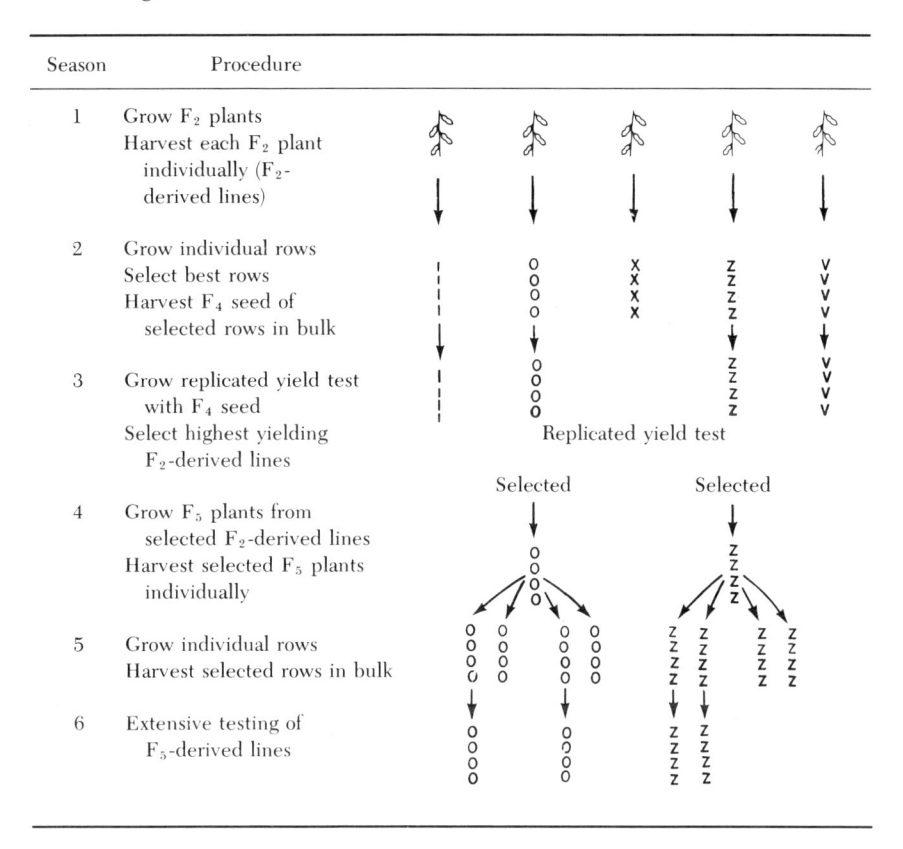

Fig. 8. Illustration of early-generation testing using F_2-derived lines.

Early-generation testing is considered a means of increasing the percentage of high-yielding pure lines that undergo extensive testing. One disadvantage of early-generation testing is that expensive yield tests are used to evaluate F_2 lines that are not sufficiently pure for use as cultivars. The same yield tests could be used for evaluating additional pure lines. Another disadvantage is that early-generation testing generally takes more time for developing a new cultivar than the single-seed descent method.

E. Recurrent Selection

Soybean breeders have begun to consider the use of a breeding system called recurrent selection to develop improved breeding populations from which superior pure lines can be selected (Brim, 1973). Early-generation testing (Fig. 9) and single-seed descent (Fig. 10) commonly are used for recurrent selection.

Season	Procedure
1	Grow plants from intermated population. Harvest each plant individually
2	Grow replicated yield test. Select highest yielding lines
3	Cross selected lines in all combinations. Harvest intermated seed
4	Grow plants from intermated seed. Repeat the procedure beginning with season 1

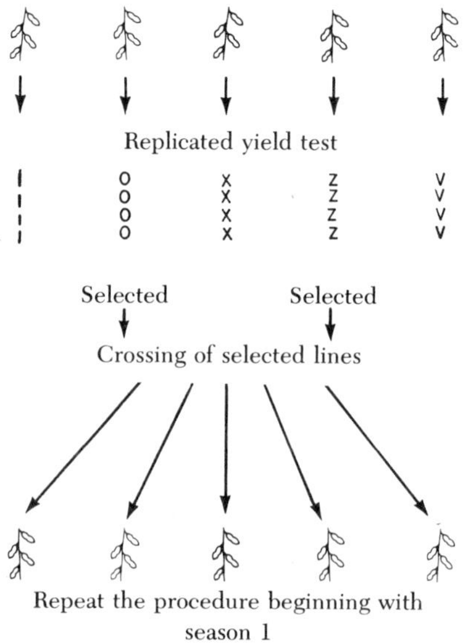

Fig. 9. Illustration of recurrent selection by early-generation testing.

The first step in recurrent selection is to develop an intermated population that includes a large number of different parents (see Fig. 4). Many parents are used in the population to increase the number of different genes that may be useful for improving a character. One technique used to form an intermated population is a stepwise approach (Fig. 4). The first step is to cross each parent to one other parent. If eight parents are used, a total of four different crosses are made. The F_1 plants from one cross are mated to F_1 plants from another cross to form two populations, then F_1 plants from the two populations are mated. The seed produced represents an intermated population that is a mixture of genes from the eight original parents.

Progeny from the population are evaluated for the character being improved and the superior ones are used as parents to form a new population (Figs. 9 and 10). Progeny from the new population are evaluated and the superior ones are crossed together to form still another new population. If selection is effective, each new population will be superior to the previous one. The process is continued until the desired level of the character is reached.

Recurrent selection can readily be integrated into a cultivar development

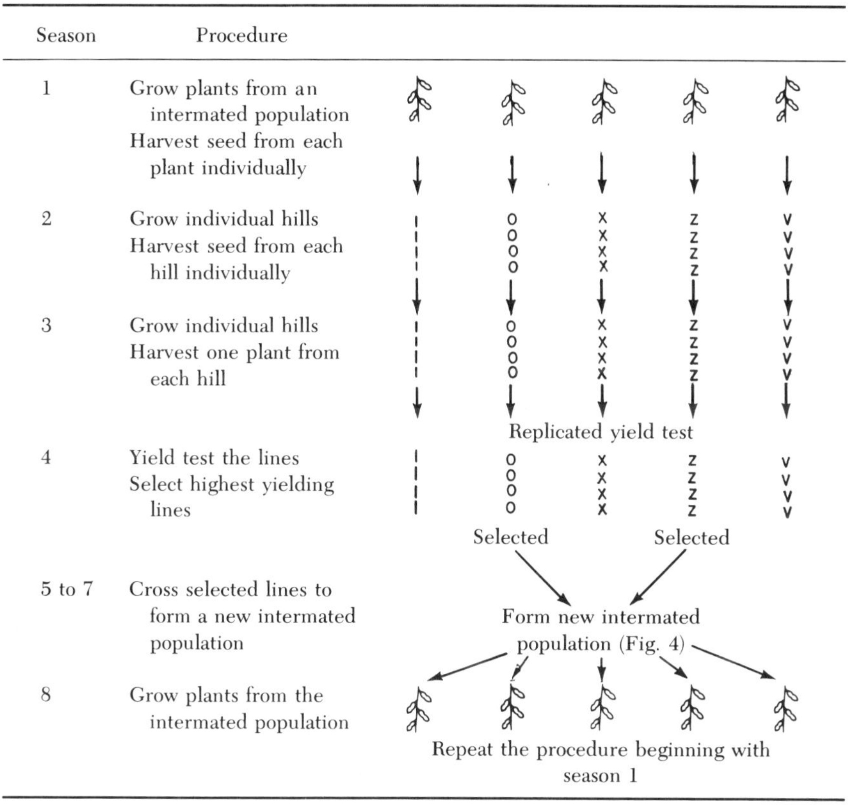
Season
Procedure
1
Grow plants from an intermated population
Harvest seed from each plant individually
2
Grow individual hills
Harvest seed from each hill individually
3
Grow individual hills
Harvest one plant from each hill
Replicated yield test
4
Yield test the lines
Select highest yielding lines
Selected
Selected
5 to 7
Cross selected lines to form a new intermated population
Form new intermated population (Fig. 4)
8
Grow plants from the intermated population
Repeat the procedure beginning with season 1

Fig. 10. Illustration of recurrent selection by testing lines developed by a single-seed descent procedure (Fig. 7).

program. The superior progeny selected as parents can be evaluated further for their potential as new cultivars.

F. Comparison of Alternative Breeding Methods

Each of the breeding methods discussed has advantages and disadvantages and the breeder must decide which method will be most efficient with the resources available. Soybean breeding has been referred to as a "numbers game" because the chance of finding a superior cultivar is improved by increasing the number of pure lines that are tested each season. Cultivar development also can be called a "time game" because the amount of improvement that can be made over a period of time is influenced by the number of years required for a cycle of selection. A cycle of selection includes the crossing of parents, development and testing of progeny to identify new parents, and the crossing of the new parents.

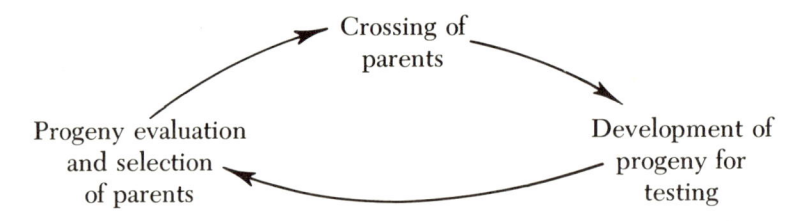

```
                    ──► Crossing of
                   ╱     parents
                  ╱             ╲
    Progeny evaluation           ╲
    and selection   ◄──        Development of
    of parents                  progeny for
                                testing
```

Time is a significant consideration in soybean breeding if maximum improvement per year is to be made in any character. For example, assume that yield is improved by 100 kg/ha for each cycle of selection. In an 18-year period, a breeder that completes a cycle every 3 years will have completed six cycles and have increased yield by 600 kg/ha. A breeder that uses 6 years per cycle will have completed three cycles and increased yield only 300 kg/ha in the same 18-year period.

Resources available for breeding have an important impact on time per cycle and the relative efficiency of breeding methods. For example, assume

Year	Season	Pedigree	Early-generation	Single-seed descent
1	Summer	F_2 plants selected	F_2 plants selected	F_2 plants grown
	Winter 1			F_3 plants grown
	Winter 2			F_4 plants grown
2	Summer	F_2 progeny rows selected	F_2 progeny rows selected	F_5 plants selected
3	Summer	F_3 progeny rows selected	Yield test of F_2 lines	F_5 progeny rows selected
	Winter 1		F_5 plants selected	
4	Summer	F_4 progeny rows selected	F_5 progeny rows selected	First yield test of F_5 lines
5	Summer	F_5 progeny rows selected	First yield test of F_5 lines	Second yield test
6	Summer	First yield test of F_5 lines	Second yield test	Third yield test

5. Breeding 143

that the breeder can grow only one crop each year. Which method would require the least time per cycle of selection: pedigree selection, early-generation testing of F_2 lines, or single-seed descent? The answer is that there is no difference among the methods when only one crop is grown each year. Seven seasons (years) are required to develop F_5 lines for testing by each method.

The methods are not the same, however, when more than one crop can be grown in a year. Assume a breeder can grow three crops each year and only the summer crop can be used for yield testing and pedigree selection because the two winter crops are in the greenhouse or winter nursery where testing is not possible. Which of the three methods require the least time per cycle of selection?

As can be seen in the tabulation shown, the single-seed descent method required the least number of years with the resources assumed in our example. Pedigree selection was least efficient because it was not possible to make visual selection during the winter, so only one crop could be grown each year. The resources available to soybean breeders vary considerably, therefore, comparisons must be made that are appropriate for each situation.

IV. BREEDING OPERATIONS

There are many operations involved in the successful use of any breeding method. Parents are crossed to obtain hybrid seed, breeding populations are grown to obtain pure lines, and decisions must be made concerning plot size, plant population, planting equipment, harvest equipment, and other factors. The breeder must select those procedures and types of equipment that permit an efficient program with the resources available.

A. Crossing Procedures

Most hybrid F_1 seed is produced by hand in a two-step operation that includes preparation of the female flower and pollination. The parts of the soybean flower considered in hybridization are the calyx, petals, stigma, and anthers (Fig. 11). When a flower bud is first visible, only the green calyx can be seen. As development progresses, the petals emerge from the calyx and unfold into a small, attractive flower. Enclosed within the petals are ten anthers (male) arranged in a circle around a pistil (female). The pistil has a stigma at its tip which receives the male sex cells (pollen) from the anthers. The stigma is receptive to pollination at least 1 day before the anthers are sufficiently developed to shed pollen. This time lag between female and

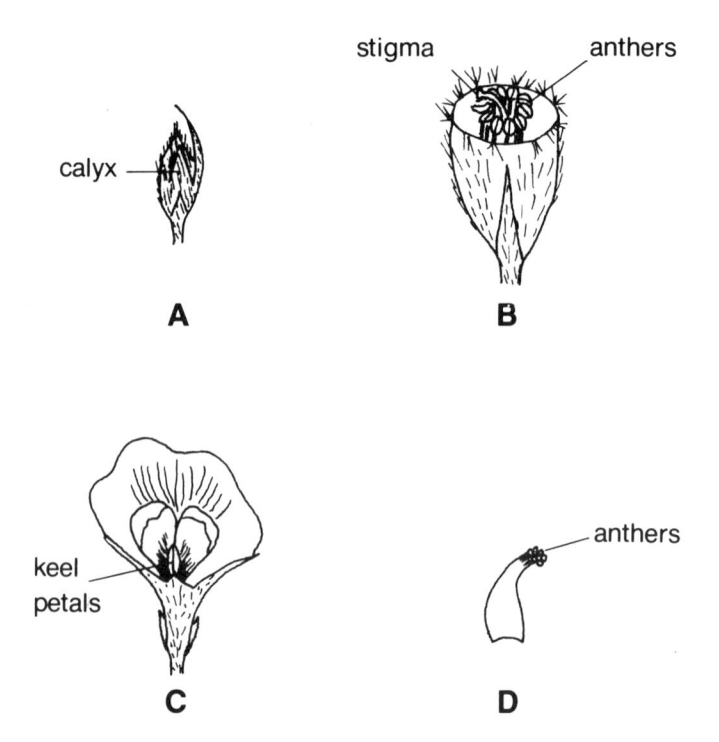

Fig. 11. Parts of the soybean flower involved in hybridization. (Drawings by Elinor L. Fehr.)

male development permits the breeder to introduce pollen from an outside source to obtain hybrid seed.

The first step in hybridization is to determine which parent will be used as the female. Whenever possible, the female parent is one that has a recessive character for which the male is dominant. White flowers, gray pubescence, tan pods, and some hilum colors commonly are used as recessive characters (Tables I and II). By using the parent with the dominant character as male, all hybrid F_1 plants will have the dominant character. Self-pollinations that were thought to be hybrids will produce plants with the recessive character of the female and can be discarded.

The next step is selection of flowers in which the stigma is receptive to pollination, but the anthers are not shedding pollen. A suitable flower is one in which the ring of anthers and the stigma are visible (Fig. 11). Female flowers generally are at the proper stage when the petals can be faintly seen through or at the top of the calyx. To prepare the flower for pollination, the five lobes of the calyx and the entire petals are removed to expose the female and male organs. Available evidence indicates that removal of the anthers is

5. Breeding

not necessary in most environments (R. L. Bernard, personal communication). Apparently an exposed stigma that is not promptly pollinated dries up and becomes nonreceptive before anthers within the flower can shed pollen.

Pollen generally is available from newly opened flowers that have a fresh appearance (Fig. 12). The golden yellow anthers are clumped together around the stigma and no anther ring is visible. Pollen is placed on the stigma of the female parent by tapping it gently with the anthers. When pollination is completed, the flower is identified with a tag.

There is considerable variation among environments in the time of day when hybridization is most successful. Some breeders must start at dawn while others have best success in the evening. In some environments, pollen is collected in the morning, dried, and used in the afternoon.

The number of crosses a person can make in a day ranges from less than 50 by inexperienced personnel to about 150 by individuals with considerable experience. The percentage of crosses that produce hybrid seed ranges from 0 to more than 75%, depending on environmental conditions and personnel.

Genetic male sterility sometimes is used to produce hybrid seed for recurrent selection programs (Brim, 1973). Fertile pollen from plants used as

Fig. 12. Hand pollination of soybeans to obtain hybrid seed.

males is carried to male sterile flowers by insects and no hand labor is needed for flower preparation or pollination.

B. Generation Advance

Advancing material from F_1 to later generations is relatively easy because soybeans are naturally self-pollinated. The breeder must decide how many pure lines it is possible to test from a population and which breeding method will be used during generation advance. Based on these decisions, one can readily calculate the amount of plant material that must be grown each generation.

The number of F_1 seeds generally is limited and the number of F_2 seed required may be high; therefore, maximum seed production per plant often is desired. F_1 plants spaced 30 cm (1 ft) apart in 75 cm (30 inch) rows can produce over 300 seeds each. Plant spacing in the F_2 and later generations depends on the breeding method used. Rows for visual selection commonly are 2 m long with a plant spacing similar to commercial plantings. The seeding rate is lower if single plants with adequate seed must be harvested.

The cultural practices adopted for generation advance are those recommended for commercial soybean production (see Chapter 6). Tests for lodging resistance, seed quality, and other characters may require special environmental conditions.

C. Yield Testing

The most expensive and time-consuming operation in soybean breeding is yield evaluation. The yield of new pure lines must be compared with existing cultivars to identify those that are superior. In establishing a yield test, the breeder must decide which lines will be compared. When a breeder has 1000 new lines to evaluate, they cannot all be grown together in one test. One large test could require several hectares of land and within that area, the productivity of the soil could vary considerably. A line that happened to be planted in a good area would yield well and a similar line planted in a less productive area would yield less. It would be difficult for the breeder, therefore, to select lines with superior genetic makeup.

Lines generally are grouped into sets of 20 or 100 entries. Lines of similar maturity frequently are put in the same set. Each set of entries is planted two or more times (replicated) at each of one or more locations (Fig. 13). A replication contains one plot for each entry in the set and entries are randomly assigned to the plots. The area required for one replication of 20–100 lines is relatively small, nevertheless, soil differences usually exist within the area. To estimate the effect of factors that are not uniform for all lines, two or

5. Breeding

6	4	8
10	2	1
7	11	5
9	3	12

Replication 1

4	6	9
12	11	3
10	5	8
7	1	2

Replication 2.

12	11	5
1	7	6
10	4	3
2	8	9

Replication 3

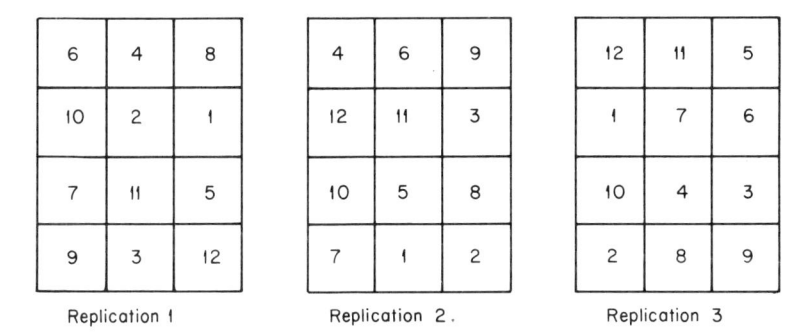

Fig. 13. Three replications of a test containing twelve cultivars. The cultivars are randomly assigned to plots within each replication.

more replications are grown. With proper statistical procedures, it then is possible to determine if differences between lines are due to genetic differences or environmental factors.

The amount of replication and number of locations increases with each successive year of yield testing to identify the most superior lines. In the first year of evaluation, the breeder may test 1000 lines in two replications at two locations. Based on that information, 200 lines may be saved for testing in two replications at four locations. Only 40 of the 200 lines may be judged superior enough for further testing in three replications at ten locations. Additional years and locations will be used to verify the performance of a line before it is released for commercial production.

D. Plot Size and Shape

The area used to grow a cultivar or line is called a plot. The size and shape of plots vary with the purpose of material, amount of seed and land available, and the type of equipment used for planting and harvesting.

The smallest plot used for testing is a single unbordered hill plot (Fig. 14). A hill is either a very short row of about 15 cm or a circle with the same diameter. The number of plants in a hill plot for yield testing generally is between six and twelve. The distance between adjacent hill plots is about 75 cm to 1 m. Bordered hill plots consisting of five or nine hills also have been used. Data are collected from the center hill to avoid any competition from adjacent plots. If some competition effects can be tolerated, all hills in a plot or the center five hills of a nine-hill plot can be harvested.

A widely used plot is a single unbordered row spaced about 1 m from adjacent plots (Fig. 15). Short rows of about 2 m are used for evaluating

HILL PLOTS

Unbordered
Single-hill

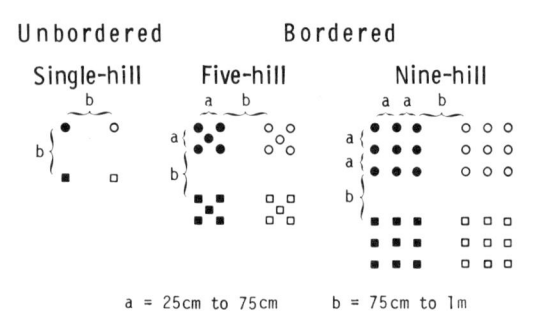

Bordered
Five-hill **Nine-hill**

a = 25cm to 75cm   b = 75cm to 1m

Fig. 14. Illustration of hill plots. Cultivars are designated by ○, ●, □, ■, and the dimensions indicated are the range of values used by breeders.

ROW PLOTS

Unbordered - Equal row spacing

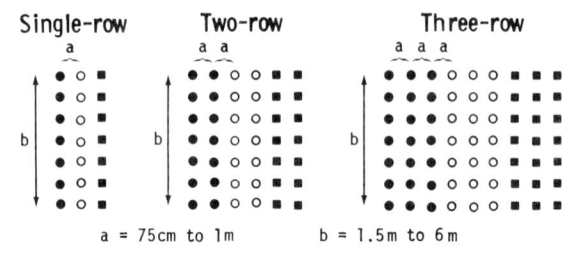

Single-row   **Two-row**   **Three-row**

a = 75cm to 1m   b = 1.5m to 6m

Unbordered - Unequal row spacing

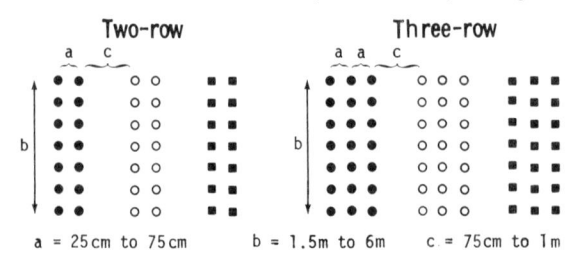

Two-row   **Three-row**

a = 25cm to 75cm   b = 1.5m to 6m   c = 75cm to 1m

Fig. 15. Illustration of unbordered plots. Cultivars are designated by ●, □, ■, and the dimensions indicated are the range of values used by breeders.

5. Breeding

lodging resistance, plant height, and similar characters, and rows about 5 m long are used for yield tests.

Multiple-row unbordered plots can be used for yield tests to reduce the influence of competition from lines in adjacent plots. Number of rows per plot is generally two or three, length of the plots 1.5–6 m, and spacing between rows 75 cm to 1 m. An unequal row spacing sometimes is used with multiple-row unbordered plots. The distance between rows within the plot is 25–75 cm, but the distance between rows of adjacent plots is 75 cm to 1 m. Narrow rows within the plot conserve land area, and the wide space between plots minimizes competition between entries.

Bordered plots are used whenever competition between adjacent lines can influence performance to an unacceptable level, particularly for yield evaluation (Fig. 16). They may be used for all yield tests in the southern United States where competition between plots can be large. In the northern United States, competition between plots can be small; therefore, some

ROW PLOTS

Bordered - Equal row spacing

Three-row
a = 50 cm to 1 m
b = 1.5 m to 6 m

Four-row
a = 50 cm to 1 m
b = 1.5 m to 6 m

Five-row
a = 15 cm to 70 cm
b = 1.5 m to 6 m

Bordered - Unequal row spacing

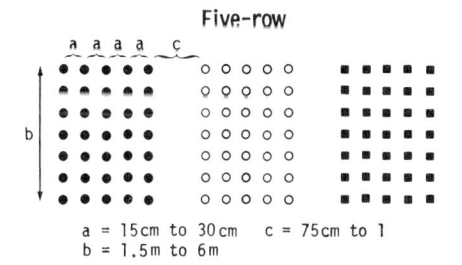

Five-row
a = 15 cm to 30 cm   c = 75 cm to 1
b = 1.5 m to 6 m

Fig. 16. Illustration of bordered row plots. Cultivars are designated by ●, ○, ■, and the dimensions indicated are the range of values used by breeders.

breeders use unbordered plots for evaluating new lines in the first and second years of testing and bordered plots thereafter. Multiple-row bordered plots generally consist of three to five rows. When narrow row spacings are used, a 75 cm to 1 m space may be left between plots to facilitate planting, cultivation, and harvest.

E. Field Operations and Equipment

1. Seedbed Preparation and Planting

Seedbed preparation for breeding plots generally is the same as used for commercial soybean production. When planting is done with hand planters, the seedbed must be more free of soil clods and crop debris than for tractor-powered planters.

Most breeding plots are planted with commercial equipment that has been adapted for use with short rows. The most common arrangement is to replace the conventional planter box with a seed distributor. The seed distributor consists of a metal cone with fins at the base. A funnel is mounted above the cone to hold the seed to be planted. As the planter reaches the beginning of the plot, the funnel is lifted and the seed falls around the base of the cone. The seed is dragged by the fins to the seed outlet, and one revolution of the cone plants one row. Row length is varied with the gears or sprockets used to change plant population for commercial plantings.

Seed dividers have been developed to simplify the planting of multiple-row plots. Seed for one plot is packaged in one envelope. The seed is poured into the divider and distributed equally to each row of the plot. Use of a seed divider reduces the number of seed envelopes needed, thereby reducing labor for both packaging and planting.

2. End-Trimming

Soybean plots are planted with a space (alley) of about 1-2 m between the end of one plot and the beginning of another. The alley separates the plots and provides a place to walk during data collection. Soybeans growing at both ends of a plot next to the alley do not have as much competition for nutrients and moisture as plants in the middle. As a result, the end plants yield more than those in the middle. The maturity of a cultivar influences how much the end plants will be affected by the alley. Early maturing cultivars are less able to benefit from the alley than late cultivars.

The end plants must be removed before row plots are harvested to accurately compare the yield potential of different cultivars. End-trimming is used to assure that all plots are of equal length and to remove the border effect of the alley. About 60 cm are removed from each end of a plot after

5. Breeding

pods on the main stem have begun to turn brown. For large breeding programs, it is not always possible to delay end-trimming until harvest, therefore, the plots may be trimmed to an equal length any time during the growing season. Special care must be taken in comparing cultivars of different maturities when plots are end-trimmed early in the season.

3. Harvesting

Self-cleaning threshers are available for harvesting individual plants or large plots. Individual plant threshers consist of a threshing cylinder and concave, and a winnowing fan. Large plot threshers, stationary and self-propelled, use both sieves and fans for seed cleaning.

When stationary plot threshers are used, each plot is identified with a tag before or immediately after it is cut. The rows are cut with a mower and the bundle is carried to the thresher. The plants are threshed, the seed is placed into a small paper or cloth bag, and the identification tag is attached to the bag.

A self-propelled plot combine reduces the labor required for plot harvest (Fig. 17). The identification tags are on the combine in the order of harvest. The combine cuts and threshes the plants, the seed drops into a bag, and the identification tag is placed on the bag. Some combines are equipped with a scale so that the seed from a plot can be weighed and recorded, then discarded if not needed for subsequent tests. A small sample from each plot may be used for moisture determination, particularly when cultivars with different maturities are in the same test. Plastic tubes also have been

Fig. 17. Self-propelled combine for harvesting row plots. (Designed by Robert C. Clark and Walter R. Fehr, and built by Allan Machine Company.)

mounted on combines to measure the volume of seed from a plot, instead of the weight.

Seed is dried before storage if the moisture content is above 15%. Drying temperatures should not exceed 40°C to maintain seed viability. Seed also is dried to obtain uniform moisture for all entries in a yield test before weighing. The duration of drying required varies with the differences among entries for moisture content and the types of drying facilities.

V. BLENDS

A soybean blend is a mixture of seed from two or more pure cultivars (Brim, 1973). A blend is not a hybrid because no hybridization is involved in preparing the seed. To produce a blend, a separate pure field is grown for each of the cultivars to be used, and the harvested seed is mixed in the desired proportion during or after cleaning.

Soybean blends were first sold as seed in the late 1960's and currently are merchandised by many seed producers in the Midwest. They became popular because they provided the seed producer with an exclusive product. Seed producers can mix together two or more cultivers and give the seed a brand name that cannot be used by any other company.

The yield of a soybean blend generally is similar to the average yield of the cultivars mixed together. Blends occasionally will outyield the average of the cultivars or even the highest yielding cultivar. The yield performance of soybean blends can be illustrated with data from studies of CORSOY and AMSOY (or AMSOY 71) mixtures. Blends of the two cultivars were tested against pure stands of each cultivar at 153 locations over the period 1969–1975 by public and private breeders (C. W. Jennings, personal communication). Many different percentages of the two cultivars were studied, but the average percentage of the two cultivars in the blends was 60% CORSOY and 40% AMSOY. The CORSOY–AMSOY blends yielded 3017 kg/ha (44.9 bushels/acre), CORSOY yielded 2977 kg/ha (44.3 bushels/acre), and AMSOY yielded 2916 kg/ha (43.4 bushels/acre). The blends outyielded both cultivars, but the increase in yield was small.

A blend can be used to minimize fluctuations in yield over different environments. Farmers may find that CORSOY is the best cultivar some years and AMSOY 71 in other years. To reduce risk, farmers can plant part of their soybean acreage to CORSOY and part to AMSOY 71. An alternative is to plant their entire acreage to a blend of the two cultivars. The blend would provide consistent performance and may give slightly higher yield than the average performance of the two cultivars grown separately.

One valuble use of blends is to provide a hedge against unpredictable production problems. Some diseases and other pests occur sporadically

5. Breeding

across locations and years. Phytophthora rot is an example of a disease that can be serious some years, but its importance is difficult to predict in advance of planting. The highest yielding cultivar available may be susceptible to such a disease, and a farmer that plants the cultivar risks a yield reduction if the disease is severe. Planting a lower yielding resistant cultivar may be safest, but would not provide the yield potential of the susceptible cultivar under disease-free conditions. A blend of the two cultivars takes partial advantage of the yield potential of the susceptible one without risking the entire crop. The CORSOY–AMSOY 71 blend has been used as a hedge against phytophthora rot, an unpredictable disease in northern Iowa. CORSOY, a susceptible cultivar, is higher yielding than the resistant cultivar, AMSOY 71. By blending the two cultivars, farmers have protection against the disease without sacrificing all the yield potential that CORSOY provides.

The disadvantage of blends for the breeder is that they must be tested to identify combinations with superior yield. The number of possible soybean blends is extremely large. For example, if ten cultivars are mixed together in all possible two-component blends, forty-five entries would result. If three different frequencies of the components were used (3:1, 1:1, 1:3) the number of entries would increase to 135. One must decide if the resources used to test blends would be better spent in the evaluation of more pure lines. Superior blends can only be improved by developing superior pure cultivars; therefore, a compromise must be made between blend and pure line testing.

The number of potential blends can be reduced by considering a cultivar's maturity, yield, and resistance to production hazards. Farmers have shown reluctance to grow a blend containing cultivars differing in maturity by more than 4 days. When early plants are ready for harvest, the farmer does not like to wait for the later component to mature. Shattering and seed quality are also potential problems when the maturity is too diverse.

Yield of cultivars in pure stand is a good indication of the potential yield of the blend. Testing blends that are mixtures of high- and low-yielding cultivars would be of little value.

Yield evaluation of blends can be minimal when they are used as a hedge against a production hazard. The primary emphasis in such circumstances is to use as little of the lower-yielding resistant cultivar as possible without sacrificing the desired protection. Yield of the blend in absence of the problem can be closely estimated with the yield of the cultivars in pure stand.

VI. HYBRIDS

There is no hybrid seed presently available for commercial production of soybeans, but a United States patent has been issued for a procedure that the inventors believe may make hybrids a reality. Production of large quantities of hybrid seed for commercial use poses some unique problems. There are four principal requirements that must be met to obtain a successful commercial hybrid: (1) the hybrid must demonstrate adequate hybrid vigor, (2) self-pollination must be avoided or minimized, (3) pollen must be transferred from the male to the female parent, and (4) cost of the hybrid seed must be consistent with its performance in comparison with pure cultivars.

Hybrid vigor is the increase in performance of the hybrid over its parents. Hybrid vigor in soybeans, as in other crops, is dependent on the parents used. Hybrids from some parents exhibit no yield increase over the parents while others have over a 15% increase. Identification of superior parental combinations requires the same type of yield testing used to develop superior pure cultivars.

There are several mechanisms that can be used to avoid or minimize self-pollination for hybrid seed production. The most widely known system is male sterility. By eliminating viable male pollen from a female plant, all seed produced on it can be hybrid. Cytoplasmic-genetic male sterility is the system used to produce hybrid seed in crops such as sorghum. Factors in the cytoplasm of cells in the female parent prevent viable pollen from being produced. The cytoplasm of the cell is passed from a female plant to all of its progeny; therefore, all the plants in rows of the female parent will be sterile. The male parent possesses a dominant gene that can override the effect of the cytoplasmic factors causing sterility. As a result, all seed harvested from female rows will be hybrid and fertile. There is no source of cytoplasmic-genetic sterility currently available in soybeans.

The system of hybrid seed production that has been patented does not utilize male sterility. Both self-pollinated and hybrid seed would be produced on the female parent. The amount of self-pollinated seed would be minimized by unique flower characteristics for the two parents. Flowers of the female parent would be open to expose the stigma, and flowers of the male parent would be open to expose the anthers. This would increase the opportunity for pollen from the male parent to reach the stigma of the female parent.

Soybean pollen is too heavy for wind transport, but is transported by insects, such as honeybees or leaf-cutter bees. Reliable hybrid seed production would require an adequate supply of the proper insects, and flowers of the parents would have to be equally attractive so the insects would pass from one parent to the other.

Hybrid seed must be produced at a price consistent with its performance in comparison with conventional cultivars. One difficulty in economical seed production is the large proportion of commercial soybean area that would be required to produce the hybrid seed. Only about 1 hectare out of every 200 of corn is used for producing hybrid seed, but in soybeans, 1 hectare out of every 18 would have to be devoted to hybrid seed production. The area requirement for hybrid seed production would be about twice that used for producing seed of conventional soybean cultivars. Providing adequate insect pollination for such a large area could pose a significant problem.

The system of hybrid seed production that has been patented is based on mechanical separation by seed size of self-pollinated and hybrid seed from the female parent. Seed size generally is determined by the genetic makeup of the female plant, not the genes of the seed, therefore, self-pollinated and hybrid seed are about the same size. The inventors have indicated, however, that in certain parent combinations the hybrid seed is 10–20% larger than the self-pollinated seed. A small-seeded female would be mated with a large-seeded male parent. Seed from the female parent would be passed over a sieve to separate the large hybrid seed from the smaller self-pollinated seed. The amount of hybrid seed per hectare that could be produced with the system would be less than with a female parent that was male-sterile and would produce only hybrid seed.

Commercial hybrid seed production remains a difficult challenge. Although the possibility of success seems relatively low, research is being conducted that may make it a reality some day.

REFERENCES

Allard, R. W. (1960). "Principles of Plant Breeding." Wiley, New York.

Bernard, R. L., and Weiss, M. G. (1973). *In* "Soybeans: Improvement, Production, and Uses" (B. E. Caldwell *et al.*, eds.), pp. 117–154. Am. Soc. Agron., Madison, Wisconsin.

Brim, C. A.(1973). *In* "Soybeans: Improvement, Production, and Uses" (B. E. Caldwell *et al.*, eds.), pp. 155–186. Am. Soc. Agron., Madison, Wisconsin.

Hartwig, E. E.(1973). *In* "Soybeans: Improvement, Production, and Uses" (B. E. Caldwell, *et al.*, eds.), pp. 187–210. Am. Soc. Agron., Madison, Wisconsin.

6

Management and Production

J. W. TANNER AND D. J. HUME

I.	Introduction	158
II.	Planting	159
	A. Soil and Climatic Requirements	159
	B. Choice of Cultivar	162
	C. Fertility	163
	D. Tillage and Seedbed Preparation	167
	E. Weed Control	168
	F. Time of Planting	170
	G. Row Width and Seeding Rate	175
	H. Planting Equipment	180
	I. Inoculation	181
	J. Seed Dressings	185
III.	Management during the Growing Season	185
	A. Selection and Use of Herbicides	185
	B. Lodging	187
	C. Diseases	188
	D. Nematodes	189
	E. Insects	201
	F. Hail Damage	201
	G. Minor Elements	207
	H. Irrigation	208
IV.	Harvesting	209
V.	Drying and Storage	214
	References	216

I. INTRODUCTION

Top production of soybeans requires the integration of climatic, edaphic, genetic, and managerial resources. Despite the fact that the American producer farms some of the world's best soils in a climatic area well-suited to soybean production, has the benefits of strong research and industrial support available, and has a marketing infra-structure second to none, it is the United States producer himself who has vaulted soybeans to the major crop status they now enjoy. In the end, it is the farmer who must match the appropriate cultivar to his climate, weather, soil, and equipment and superimpose his overall managerial ability on the system to produce top yields.

Within an area, the major difference in yield among farmers is their management ability. With the same resources available, farm yields within an area will range from much below average to much above average and occasionally to near-record. The farmers who consistently produce the high yields are those who select the proper inputs and are timely in all aspects of their operation. In general, these are the farmers who follow the recommendations provided by their local advisory personnel, recommendations based on the results of sound research findings. Certainly nothing could increase average soybean yields quicker than a broader acceptance and application of the knowledge already available.

In the United States, soybeans have traditionally been treated as a secondary crop and, as such, have been produced on the "remaining acreage," utilizing equipment designed for other crops. In recent years, the crop has been recognized as a first-choice crop by many farmers and, as yields per hectare and total area under production have grown, equipment and systems unique to the soybean crop have been designed and continue to develop.

Technological advances in soybean research have lagged behind those in other major crops, not as a result of lack of quality of research, but rather as a result of lack of quantity (Nelson, 1976). Until relatively recently the total research effort devoted to soybeans was only a small proportion of that devoted to corn, wheat, or cotton. As the demand for the crop continued to expand during the 1960's, both state and federal research institutions greatly expanded their research teams. More recently, the overall effort in breeding new cultivars has increased dramatically as a result of the entry of many commercial seed companies into the production and marketing of their own lines. Undoubtedly, the development of hybrid soybeans could result in an explosion of new cultivars of relatively short lifespan, similar to the current situation with hybrid maize.

The effort devoted to soybean research in other countries is increasing dramatically. There is considerable interest in soybeans in tropical countries for food production. In temperate regions, Brazil, Australia, eastern and western Europe, the U.S.S.R., China, and Japan have all mounted significant research programs. As long as a free flow of information and germ plasm among all countries continues, the rate of improvement in the crop can only accelerate.

Production practices described herein are essentially those employed in the United States. Soybean management systems in other parts of the world, particularly the tropics, may vary considerably from these described. Such differences may arise because of differences in environmental effects, social uniqueness, or government policy. Although not directly applicable to certain areas of the world, the United States systems may at least present successful examples.

II. PLANTING

A. Soil and Climatic Requirements

Soybeans are produced on a broad range of well-drained soil types; however, less stability for production exists on sandy soils. Difficulties in planting and emergence are often encountered on the heavier clay soils although, once established, soybeans are better adapted to these soils than most other crops. Medium-textured soils appear to be ideal for soybean production. Soybeans also produce well on high organic or muck soils, provided nutrient elements are not deficient. Soils low in fertility should be augmented with the appropriate nutrients, as determined by soil or tissue test. The soil pH should be in the range of 6.0–6.8 for good nodulation and plant growth. Dry soils or wet soils adversely affect both germination and growth in soybeans. Soybean growth, development, and yield are reduced by even moderate amounts of soil compaction; fields where compaction is a problem should be avoided unless special tillage practices can be employed to offset the effects.

Few crops are as flexible within a farm rotation as are soybeans. Because soybeans are adapted to well-drained, medium-textured soils, they are most often found in rotation with other crops adapted to similar soils, including corn and small grains in the north and corn and cotton in the south. Soybeans are not quite as drought-tolerant as corn and cotton, and thus they appear less frequently than these crops on the droughtier soils. More frequently they tend to show up on some of the wetter soils or bottom lands which must be worked and planted later in the spring. Although soybeans may appear in a rotation which includes peanuts or tobacco, they should follow these crops rather than precede them in order to avoid carry over of insect and/or disease pests.

In the United States midwest, soybeans usually follow corn or small grains because soybeans are efficient in utilizing the residual fertilizer applied to these crops. When a legume sod is in the rotation, corn, as a nonlegume, can much more efficiently utilize the nitrogen contained in the sod residue, which would tend to reduce nitrogen fixation by the soybeans. Similarly, in a corn–soybean rotation, the corn can benefit from the nitrogen residues produced by the soybean. Although soybeans have been grown with success continuously in some fields (Miller, 1966), the practice is not recommended because of the possible buildup of specific weeds, diseases, and insect pests, and the tendency toward compaction in some soils.

There are several advantages to including soybeans in a rotation: (1) certain weeds which may be difficult to control in other crops may be controlled either chemically or by cultivation in soybeans; (2) because soybeans can be planted and harvested over an extended period, they complement the work distribution of many other crops; (3) they perform better than most other crops after spring plowing; (4) soybean herbicides do not leave toxic residues for the following crop; (5) soybeans in a rotation are not as susceptible to attack by soybean pests which may tend to build up under continuous production; (6) soybeans in a rotation may break disease or insect cycles of pests which attack the other crops in the rotation; (7) soybeans are thought to improve the physical condition of the soil. They undoubtedly are no more severe on the soil than corn or small grains (Calland, 1949). Their dense canopy protects the soil from rain, sun, and wind and most farmers believe that they leave the soil in good tilth, making subsequent tillage easy; (8) soybean straw and roots add 50–70 kg/ha of organic nitrogen to the soils; (9) soybeans can be grown as an emergency hay or silage crop in the event of other crop failures; (10) grown as a plow-down crop they represent an excellent source of green manure.

The climatic requirements for soybeans are similar to those for corn. Cultivars have been developed which are adapted over latitudes ranging from the equator to as far north as Sweden. Of necessity, the northern types require less time to reach maturity because of the shorter growing season available in those regions.

Essentially all stages of growth are affected by temperature. Soybeans germinate and emerge rapidly at soil temperatures of 30°C. Temperatures of 20°C delay germination and emergence and at 10°C germination and emergence are very slow. Cultivars differ significantly in their ability to germinate and emerge at these low temperatures. These differences appear to be due to differences in cold tolerance per se and differences in hypocotyl elongation (Littlejohns and Tanner, 1976). Under the cool soil conditions encountered in early planting, emergence may take 2–3 weeks whereas under later, warmer conditions, emergence may occur in 3–5 days. Cool soil

6. Management and Production

conditions, which result in delayed emergence, expose the germinating seed to increased risk of damage from soil pests and diseases which frequently cause reductions in stands. As a result, except in the short season areas, soybeans are not normally planted until soil temperatures are moderately warm.

Temperature also has a great effect on rates of growth, development, and photosynthesis (dry matter accumulation). Growth, as measured by height and leaf expansion, occurs more rapidly under warm conditions. Under more rapid conditions of growth, ground cover and weed suppression are achieved earlier but the crop has a greater tendency to lodge.

Shorter plants may result from too early or too late planting. In the north early plantings are slow to emerge and develop slowly under the cool conditions prevalent at that time; in the south early planted soybeans may be photo-induced by the short, but increasing day lengths of the spring. Late planted soybeans, except those which are insensitive to photoperiod, are photo-induced by the shorter photoperiods before they achieve their normal stature.

The minimum temperature for most growth processes in soybeans appears to be about 10°C. High temperatures (over 35°C) have been shown to reduce growth of internodes. Temperature effects on rate of development, particularly time to flowering, have been shown by many investigators (Van Schaik and Probst, 1958; Huxley and Summerfield, 1974; Major et al., 1975a,b). Numerous authors have attempted to predict flowering date by means of various temperature accumulation systems. However, because of the interaction with photoperiod, such systems have limited value except within small latitudinal bands. In most cases 10°C is taken as a minimum temperature for soybean development. Temperatures below 25°C delay both flowering and maturity; high temperatures, over 35°C, appear to retard development. Considerable genetic differences exist among cultivars with regard to the effect of both day and night temperatures on flowering (Tanner and Hume, 1976).

In North America soybean cultivars appear to be adapted to specific latitudinal (day length) zones (see Chapter 2 Section II, C and Chapter 5 Section II, A). The relative maturity of a soybean cultivar within a certain zone places it within a specific Maturity Group. Officially, there are ten Maturity Groups ranging from 00 (earliest) to VIII (latest). In fact, because even earlier and later types exist, the maturity variability among cultivars should show a spread of 0000 through to Group X. Because of the close relationship between day length and temperature, the system of classification is adequate within the soybean growing area of North America. However, the classification system breaks down in other parts of the world where, because of altitude and seasonal differences, cultivars may not fit their predicted Maturity Grouping. Such problems are of no concern to a farmer who makes a selection from a list of cultivars recommended for his area, normally only described as being early, medium, or late in maturity.

B. Choice of Cultivar

Soybean researchers classify cultivars by Maturity Group (00 through VIII). Groups 00, 0, I, and II are adapted to the longer day length areas of southern Canada and the United States midwest. Groups III–VIII are adapted further south with Group VIII being the latest Group grown extensively in the continental United States. There is a maturity difference of 10–15 days, depending on year and location, between successive Maturity Groups. By far the largest area in the United States is planted to Group II soybeans, which are adapted to the extensive production areas of the midwest (Hartwig, 1973). Grown north of its area of adaptation a soybean will be more vegetative and flower and mature later; grown south of its area of adaptation it will flower earlier, produce less vegetative growth and mature earlier.

In choosing a cultivar, a farmer's first decision should be to grow only recommended cultivars. Local government research stations devote a considerable amount of their time to evaluating cultivars at several locations over the area they service. Although such tests may not isolate one superior cultivar, the data collected and reported are useful in identifying obvious weaknesses in a cultivar, those poorly adapted, and the inferior cultivars. Any cultivar grown which is not recommended should be restricted to a small observational planting.

Only good quality seed should be selected. The upper limit of yield is determined when the seed is planted; if poor quality seed is used top yields cannot be expected even if growing conditions are excellent. The seed should contain few splits and/or cracks, seed coats should be intact, and seed should be free from weeds, other seeds, and trash. Seed germination should be 85% or greater, and preferably near 95%. In northern areas or for early planting, the seed response in cold germination tests may provide additional, useful information (Searle, 1975). Seed of uniform size is important for a good distribution of seed within the row during planting. Seeds of uniform size produce more uniform plants which tend to compete more evenly with adjacent plants, resulting in fewer stunted, barren plants. The use of certified seed is the best assurance that the trueness to cultivar and the quality criteria will be met. Because soybeans are self-pollinated, harvested seed can be saved and grown in subsequent years. However, even though the genetic integrity of the seed is not at issue, a farmer saving seed should ensure that his seed is cleaned and sized by screening. Prior to planting he should conduct a germination test in order that he may adjust the seeding rate to compensate for any reduction in the germination percentage of his seed.

In selecting a cultivar, the first consideration should be maturity. An earlier-than-adapted cultivar will have a reduced yield; a later-than-adapted cultivar may not mature and will produce a low yield of poor quality seed. Within an area there are usually early-, medium-, and late-adapted lines. In general the full-season cultivars produce the highest yield. However, the early-adapted cultivars, particularly if planted early in narrow row widths, may produce yields equivalent to full-season cultivars.

From those cultivars adapted to an area, those with high yields should be selected. The high-yielding cultivars on a recommended list have demonstrated their stability and reliability of performance over several locations and years. In some areas, diseases, insects and/or nematodes may represent a serious problem. In many cases, cultivars resistant to these may be available. In areas where specific diseases or pests are a problem, resistance to these may be more important than top yields. In areas of good rainfall, lodging may be a serious problem. Selection of cultivars less susceptible to lodging is essential in such problem areas. Unfortunately, the environmental conditions which tend to produce top yields (high fertility, good moisture) are those which frequently promote lodging (Cooper, 1971). Choice of lodging-resistant cultivars and other agronomic practices can reduce this risk.

C. Fertility

1. General

When soybeans were first extensively planted in the corn belt, it was generally believed that there was essentially no response to applied fertility and that soybeans responded to residual fertility. In part this conclusion was drawn as a result of the lower yield levels being obtained at the time because of the cultivars used and difficulties in controlling weeds and, in part because, from a seed energy content, one could not expect the same bushel per acre response from soybeans as from corn. Corn yield increases of 100–600 kg/ha from phosphorus or potash application could be shown to be a significant increase; soybean yield increases of 100–400 kg/ha to the same amounts of applied fertilizer seldom showed a statistical significance even though they were consistent. As agronomic practices were improved and soybean yield levels increased, yield responses became easier to demonstrate experimentally and it is generally accepted now that where the soil fertility is such that applied fertilizer would result in a yield increase in corn or other crops, a yield response would be achieved on soybeans as well (Bray, 1961; Voss, 1967).

It has been generally observed for years in the United States midwest that high soybean yields tend to be produced in fields that produce high yields of corn. In the main this is a reflection of the overall managerial ability of the farmer; more specifically, it is a result of a carryover of high fertility and good weed control from the preceding corn crop.

A fertility program for soybeans should start with a soil test. Fields with low to medium levels of phosphate or potassium should receive fertilizer and should show a response. Most soybean fields receive little, if any, direct fertilizer application, as most farmers still prefer to apply the bulk of their fertilizer to other, apparently more responsive, crops in the rotation. Certainly the establishment of a high level of fertility in the soil, with continuing maintenance levels of fertilizer, is the best system for all crops. Fertility levels can be monitored by soil testing and/or tissue testing, if the services are available.

Fertilizers may be applied broadcast, banded, with the seed ("pop-up"), or foliarly. Broadcast applications are recommended where the amounts required are fairly high or where narrow row production makes it inefficient to apply in the band. Broadcast applications can be made at any convenient time during the year but the soil should be worked well prior to planting to ensure good distribution of the nutrients at levels below the surface. Small amounts of fertilizer are frequently banded during planting. To ensure that burning does not occur, most researchers recommend that the band be at least 5 cm beside and 5 cm below the seed. Where fertility levels are medium to high, banding probably represents the most efficient use of the fertilizer applied. Where wide row production is employed the root system may not efficiently forage the between-row area for nutrients; banding eliminates this inefficiency. Yield responses of soybeans to "pop-up" fertilizers have been infrequent and, because the risk of burning the seed is greater with soybeans than other crops, the use of "pop-up" fertilizers is not recommended (Clapp and Small, 1970). With the exception of some recent studies on foliar applications of major nutrient elements (Garcia L. and Hanway, 1976), there has been little evidence to support foliar feeding. However, foliar applications of minor elements, when deficiency symptoms become evident, represent the simplest, most reliable, and cheapest means of remedying the problem.

2. Nitrogen

When properly nodulated, soybeans derive much of their nitrogen through the process of symbiotic nitrogen fixation. The proportion of the plant nitrogen derived from nodule fixation depends largely on the nitrogen level of the soil. The lower the level of nitrate and ammonium in the soil, the greater the fixation. Most research reports from the midwest estimate that symbiotic fixation produces 40–70% of the crop requirement; more southern locations, with presumably lower levels of soil nitrogen, show higher proportions of the total plant nitrogen derived symbiotically.

As a general practice, nitrogen fertilization of soybeans is not recommended, even where soil nitrogen levels are low. Many farmers use small amounts of nitrogen on their starter fertilizer (15–20 kg/ha). The value of this starter nitrogen to final yield may be difficult to document but, particularly if the soil is cool at planting time, this readily available nitrogen may assist in promoting growth during the period prior to the onset of symbiotic fixation. The practice tends to be more prevalent in the northern areas of production, presumably because the soils are cooler.

The symbiotic bacteria, *Rhizobium japonicum*, enter the root during the first few days after germination (see Chapter 2 Sections I, D and II, E). Nodules start to form 12–14 days after planting, and nitrogen fixation, as evidenced by greener plant color, commences 18–21 days after planting (earlier or later depending on temperature). Nitrogen fixation increases as the size of the plant increases and reaches a peak shortly after the plant becomes fully reproductive (Hardy *et al.*, 1968). The bacteria in the nodules depend upon the plant for their energy supply. Prior to the reproductive phase of growth, the nodules can compete as a carbohydrate sink. However, once the plant is fully reproductive, the seeds act as a much stronger sink for carbohydrates than do the nodules and the nodules decrease in nitrogen fixing activity and eventually senesce (Hume and Criswell, 1973).

Numerous experiments have been conducted, with few successes, attempting to demonstrate a yield response from applied nitrogen fertilizer. Invariably as nitrogen fertility is increased, the resulting reduction in nodulation and fixation occurs proportionally, with the result being no yield increase. Even when nitrogen is applied at levels sufficient to greatly inhibit nodulation, yields have remained about the same as those from a well-nodulated crop grown without added nitrogen fertilizer (Weber, 1966; Welch *et al.*, 1973). It appears that as long as a plant can meet its nitrogen requirements from the soil, its main source of nitrogen will be from the soil; when soil nitrogen levels are low, the nodules form and symbiotic nitrogen fixation becomes increasingly important. Nitrogen fertilization does not augment the symbiotic system; it replaces it, with the net result being no response.

Nitrogen fertilization, while seldom showing a yield response, frequently results in visual responses such as darker green foliage color and increased vegetative growth and height. The latter response may lead to increased lodging and yield reductions.

There have been several attempts made to negate the inhibitory effect of fertilizer nitrogen on nodulation. Deep placement of nitrogen fertilizers below the nodule zone has shown occasional increases in yield (Criswell et al., 1976). Other researchers have investigated the use of slow-release fertilizers, organic sources of nitrogen, and nitrogen stabilizers, such as nitrapyrin, all with inconsistent results. Others are seeking strains of *Rhizobium* which can continue to fix nitrogen in the presence of high available soil nitrogen. Another approach has been to develop a plant type which will not produce excess vegetative growth and remain short and lodging-resistant when provided with high nitrogen, either symbiotically-produced or applied (Cooper, 1974). There is renewed interest, too, in foliar application of nitrogen and other nutrients, in an attempt to keep the metabolic factory of the plant operating actively, and thus extending the grain-filling period by delaying senescence (Garcia L. and Hanway, 1976).

Although the current recommendation that no nitrogen need be applied to well-nodulated soybeans is generally accepted, many researchers believe the true picture on nitrogen fertilization has not yet emerged. The amount of nitrogen contained in a 3600 kg/ha soybean crop is in excess of 300 kg; seldom has nitrogen fixation been measured at levels sufficiently high to account for this amount. Obviously soil nitrogen is an important source of nitrogen in high-yielding, well-nodulated soybean plantings. There have been a sufficient number of experiments showing responses to nitrogen fertilization that they cannot be discounted or ignored. The renewed interest in foliar applications during pod filling demonstrates this belief. However elusive the solution might be, in time it will be found.

3. Lime

Proper soil pH is an essential input for high soybean production. Soil pH in the range of 6.2–6.8 is assumed by most researchers to be optimum for the crop. The range may be slightly higher or lower on specific soil types. Situations of low pH are a much more common problem than situations with a high pH, although the latter problem exists on certain high lime soils.

The addition of lime to soils where the pH is below the optimum range has several beneficial effects: nutrient availability is increased; there is reduced risk of toxicity from manganese and aluminum; nodulation and nitrogen fixation are increased; and in many soils the tilth is improved.

The most common problem associated with low pH is molybdenum deficiency. Molybdenum is not only an essential plant element, but is important specifically in the nitrogen fixation process. At lower pH levels, liming can remedy the deficiency. At threshold levels of deficiency the use of a molydenum seed dressing can show dramatic results. This interaction between molydenum and nitrogen fixation has been convincingly demonstrated by means of isogenic nodulating and nonnodulating soybean lines (Parker and Harris, 1977).

At high pH levels, depending upon soil type, iron, zinc, manganese, or copper may become deficient. When iron deficiency symptoms appear (interveinal yellowing, chlorosis), foliar applications of ferrous sulfate or chelated iron compounds are used to rectify the deficiency. Cultivars differ considerably in their tolerance to iron deficiency and in those areas where calcareous soils consistently produce the problem, e.g., southern Minnesota and northeastern Iowa, plant breeders are working to incorporate tolerance to the deficiency into their lines.

Manganese deficiency, when it is a recurring problem, is usually rectified by the application of manganese sulfate or oxide in the banded fertilizer or foliarly applied manganese sulfate. The foliar method may be used when symptoms show up unexpectedly. The deficient element is sometimes applied in mixture with pesticides.

Copper and zinc deficiencies are less frequent problems. There is increased risk of copper deficiency on high organic matter soils and some sands. Deficiency of either zinc or copper can be eliminated by means of either row-applied or foliar formulations of the nutrient concerned.

D. Tillage and Seedbed Preparation

The choice of the system of tillage employed to prepare a seedbed depends upon several factors. The effect on yield has been of prime importance. With the rapidly increasing price of energy, the cost of operating machinery is becoming an increasingly important factor. Because timeliness of operation is often critical, the time devoted to tillage should not be such as to delay planting and/or weed control operations. The type of soil being tilled is a major factor in the selection of both the equipment used and the overall system. Lastly, the equipment available, which is frequently related to other crops grown, may determine, in part, the system of tillage employed.

The ultimate objective of tillage is the preparation of a good seedbed. Traditionally a good seedbed was firm below and smooth or level enough to permit the planting equipment to place the seed at a uniform, desired depth; the surface soil was fine enough to assure good seed-soil contact, essential for rapid germination and emergence. A smooth, friable, clod-free seedbed is also essential for the effective use of preemergence herbicides. Overworking the soil should be avoided as, with excess rain, the soil may puddle, leading to poor infiltration and high runoff of water and soil. Subsequently air and water movement into the soil may be restricted and crusting may occur, resulting in poor emergence and growth. After plowing, the least tillage possible to establish a good seedbed should be employed in order to avoid overworking and reduce the risk of compaction (Nelson *et al.*, 1975).

Good seedbeds can be established by various means. Conventionally soybean land was moldboard plowed in the fall, disked and harrowed in the spring, and planted. This basic system has been modified with the increasing use of disk plows, chisel plows, disk chisels and disks, depending on the soil type, erosion hazard, and time available. Most systems, apart from no-till planting, can produce a seedbed capable of producing top soybean yields, although not all systems will work on every soil type.

There is renewed interest in many soybean growing areas in conservation tillage. The prime objective in conservation tillage is the prevention of wind and water erosion of soil. In addition to providing an adequate seedbed, conservation tillage attempts to (1) protect the soil for as much of the year as possible with either crop residue or crop, (2) produce a roughness of surface sufficient to increase water infiltration and reduce runoff of water and soil; (3) divide long slopes into short slopes by means of terraces and/or strip crops; and (4) plant natural waterways to grass to prevent gully development. These objectives can be met and still produce an adequate seedbed. The key to conservation tillage is avoidance of conventional fall moldboard plowing in favor of those operations which do not bury the vegetation but retain considerable trash on the surface and leave a rough, porous surface to increase infiltration.

Conservation tillage can visibly reduce water and soil loss. In dry years the additional water stored in the soil can provide the stability required to produce a good crop. However, because the soil is not turned over as by conventional means, there are increased problems with weed seed germination and greater risk of disease and insect carryover. The surface roughness and trash also make planting more difficult and the attainment of good stands less likely unless slightly higher seeding rates are used. In addition, the surface roughness and trash reduce the effectiveness of preemergence herbicides at recommended rates.

The tillage system finally employed must be an individual decision based on the combination of considerations facing the farmer. Certainly experience and understanding the limitations of the soil are the prime factors in the decision as to the system to be employed. However, whatever system is used, the achievement of a soil environment suitable for rapid emergence and subsequent good growth is the essential first step towards top soybean yields.

E. Weed Control

Weeds continue to be a major deterrent in reaching the yield potential inherent in the crop. The yield reductions attributable to weeds have been well documented. A considerable proportion of the tillage requirements for producing a crop is directed against weeds. Undoubtedly the main deterrent in shifting from the traditional midwestern row culture of soybeans to drilled or broadcast plantings has been the uncertainty of controlling weeds without interrow cultivation. However, the outlook for complete chemical control has never looked better.

The weed crop in any year is an inheritance from poor weed control in previous years. Weeds tend to be a contained population and the major weed problems that arise are a reflection of the management history of the field. Because weeds do not arise by spontaneous generation, the care taken in ensuring that weeds do not go to seed, that harvesting equipment is not transporting weed seeds, and that clean seed is used for all crops in the rotation, is an integral part of an overall weed program.

Proper tillage systems are the first operational prerequisite in reducing weed problems. Primary tillage operations which bury as great a proportion of weed seeds at lower depths in the soil as possible can reduce the total weed population which will germinate. The crop should be planted as soon as possible after the final seedbed preparation so that the soybean crop is not in competition with more advanced weed plants. Until recently, the most important single piece of field equipment used in soybeans was the rotary hoe. Quite apart from its value in assisting emergence in crusted soils, the rotary hoe represented the most reliable means of controlling early weed growth and competition. Subsequently, interrow cultivation was practiced primarily for weed control.

Often competition from vigorously growing soybeans represents a major contribution to midseason weed control. For many weed species the dense canopy produced by the soybean crop is sufficient to repress their competitive effect on the crop. A few of the more vigorous weeds, however, grow up through the canopy, reduce yields and add to the weed seed return to the soil.

During the past 20 years phenomenal progress has been made in the battle to control weeds by chemicals. Reliable chemical weed control not only reduces tillage operations performed only to control weeds, but, more importantly, can free the soybean crop from the traditional wide row culture required to accomplish mechanical weed control. As chemical control of weeds becomes more successful, more plantings of soybeans are being made with narrower rows. The full impact of this system change will not be felt until cultivars adapted to this new narrow row culture are developed.

Today's soybean producer has increasing flexibility in herbicide choice. The most reliable chemicals are those which are applied before the beans emerge. They may be preplant-incorporated or preemergence chemicals. These represent the backbone of a weed control program. Herbicides are also available for use just at plant emergence or cracking time. These herbicides might be described as a chemical rotary hoe. More recently the herbicide arsenal has been expanded with the development and release of selective postemergence chemicals. These chemicals cannot accomplish complete weed control in themselves but do present farmers, particularly those unable to cultivate narrow row soybeans, with a rescue operation that can accomplish the adequate level of weed control necessary for high yields.

Frequently weeds difficult to control in soybeans can more easily be controlled with herbicides in the other crops in the rotation. Although some of the former reasons for employing a rotation no longer apply, the value of rotations in insect, disease, and weed control still exists. Rotation of the herbicides used in a rotation of crops may be as important as rotation of the crops themselves.

As indicated earlier, the crop itself, as the canopy develops, represents an important aspect of weed control. Certainly prior to the development of chemicals for weed control, large, vigorous, leafy soybean plants, planted in 100-cm rows, were better able to compete with weeds than shorter plants. Such bushy plants, either consciously or unconsciously, emerged out of breeding programs. So long as weeds were a problem, there was an advantage for this type. In recent years, with greater dependability of herbicides, cultivars designed more with attributes that contribute to yield per se have been developed. The so-called "thin lines" and, more recently the semi-dwarf determinate types, are showing that, placed in the proper narrow row cultural environment for yield, these new plant types can raise the yield potential of soybeans to new heights.

F. Time of Planting

1. Conventional

The response of a soybean plant to planting date depends upon the environmental conditions subsequent to planting. The three most important environmental factors involved are temperature, day length (photoperiod), and moisture distribution.

In the northern areas of production in the United States, the cultivars grown are less sensitive to day length, with some cultivars being essentially insensitive (Criswell and Hume, 1972). In these regions temperature accumulation indexes (heat unit systems) can be used to predict flowering date and maturity. Further south both temperature and day length affect flowering and maturity. In the southern part of the United States, temperature is minor in its effect compared to photoperiod. Under tropical conditions, rainfall pattern and distribution become the prime factors in determining planting date, and planting usually is made at the onset of the rainy season.

As a general rule, full-season cultivars should be planted as soon as the soil temperature or day length will permit. It is interesting to note that the period May 15–20 would be included in almost all state planting date recommendations, although it should be emphasized that, from north to south, there is a great difference in photoperiodic responsiveness of the cultivars.

In the midwestern United States, mid-May plantings have been generally recommended for full-season cultivars. Recent experiments have indicated some yield response to slightly earlier plantings (Ryder and Beuerlein, 1977). Soybeans have traditionally been planted after corn has been planted. In recent years, because of the increased emphasis on earlier dates of corn planting, soybean planting dates have moved earlier as well. The response to early planting in corn has been shown to be about 60 kg/ha per day; in soybeans the response is about 20–30 kg/ha per day.

In the northern areas of soybean production, mid-May soil temperatures may retard germination and emergence; seed planted prior to mid-May could take 2–3 weeks to emerge; early June plantings might emerge in 3–5 days. Germination at 10°C soil temperature is slow and the seed is predisposed to attack by soil insects and diseases. Seed treatment with insecticide and fungicide is essential. Early planting may also expose the seedling to the risk of late spring frosts. Corn seedings frozen off are seldom permanently injured as only leaf tissue is destroyed; the apical growing point, being below the soil surface, continues to develop. Because of the epigeal nature of emergence of soybeans, the apical growing point and the axillary buds in the cotyledon node are all exposed to injury from an early frost. When such frosts occur, there is no meristematic area from which new buds can arise and the plant dies. However, the general observation has been that soybeans can withstand, without visible injury, temperatures which cause frost damage to corn.

In the corn belt, a delay in planting of a full-season adapted cultivar reduces yield more than it delays maturity. As a result of the sensitivity to photoperiod, a delay in planting date will not affect flowering date proportionally. Because the plant does not have as much time to develop vegetatively before flowering, height and leaf area, and ultimately yield, will be reduced and maturity will be delayed somewhat. In order to compensate for reductions in height and leaf area which arise from late planting, it has been generally recommended that later planted beans be grown in narrower rows. Early-adapted lines may yield better when planted slightly later than when planted early. When planting is delayed beyond early June, the latest cultivar which can still mature when planted at that date should be grown. Moderate delays in planting may result in taller plants and increased lodging. Early to mid-June plantings are shorter, less prone to lodging, with low bottom pod height and reduced yields.

In the southern United States, despite the fact that soil temperatures are adequate, plantings are normally delayed until mid-May in order to meet the proper day length conditions for the southern type soybeans, which flower on shorter days than northern types. If early planting takes place, it is possible to have sufficient growth for the plants to be photo-induced by the short, but increasing, day lengths of the spring. When early planting results in this premature flowering, plant size and yield will be reduced. Under mid-May plantings, the plants will not flower until induced by shorter photoperiods in July. Plantings made at this time will produce normal, full-sized plants with good yields. Later plantings will be photo-induced before adequate vegetative growth is achieved, resulting in shorter plants with reduced yields.

In the south, full-season beans yield better than mid- or short-season beans at early, normal, or late plantings. Later plantings require late-maturing cultivars in order to achieve sufficient vegetative growth to produce a good yield.

2. Double Cropping

Double cropping soybeans is the extreme situation in late planting. Practiced in more southern regions for several years, the system is spreading north. This spread is made possible by improved technology and driven by economics. The earlier the soybean crop can be planted, the more "normal" the plants will be, and the better the odds for success. The later the planting, the shorter the plants, the lower the pod height, and the greater the risk that the crop will not mature. Success depends more on soil moisture availability than any other factor and the practices recommended are geared to give as much stability as possible to this factor.

As a general rule, soil moisture levels decline as the season progresses. Earlier planted crops have a better chance of germinating and growing than those planted later. In order to improve the odds of succeeding, earlier maturing cultivars of small grains are available for use in double crop situations. As an alternative, some farmers remove the small grains earlier by swathing prior to combining or by harvesting the grain at higher moisture levels (20–25%) and drying the crop. The few days gained in planting date can result in sufficient additional moisture to ensure a crop in a dry year.

The soybean crop should be planted as soon as possible after the small grain crop is removed. Conventional tillage practices result in delayed planting and increased moisture loss. As a result, the no-till system of planting directly into grain stubble has been demonstrated to be the most reliable means of establishing the crop.

Although no-till planters vary, they are comprised of three basic components: a rolling coulter, fluted or waffled, to cut the surface vegetation and prepare the mini seedbed; a double disk opener to open the soil for seed

6. Management and Production

Fig. 1. A no-till planter operating in cereal grain stubble. Press wheels are shown at the back of the planter. Details of the fluted coulter and double disk openers are shown in the insert. Photographs courtesy of Allis-Chalmers Corp.

placement; and a press wheel, frequently with a single ridge, to firm the soil around the seed (Fig. 1).

Proper straw management can greatly reduce planting problems. Removal of as much straw as possible will make planting easier, ensure better placement and coverage of seed, and improve weed control, as less of the herbicide is applied to the straw and more to existing weeds and to the soil surface. Where straw removal is not practiced, a straw chopper, leaving a stubble height of 20–25 cm, should be employed to produce a soil mulch, with a good, even distribution of straw being required. No-till planting tends to increase lower pod height, important at harvest time, reduces risk of water or wind erosion, and provides a firm, level field which aids in harvesting.

In areas where environmental regulations permit, burning of straw is common. Although burning reduces planting problems and permits easier weed control, the loss of organic matter inherent in the practice is viewed, at best, as a necessary evil.

Weed control under the no-tillage system depends entirely on chemicals. Most often recommended is a tank mix of a contact, burnoff chemical, such as paraquat, and a preemergent residual herbicide such as linuron or linuron + alachlor. Problems in weed control arise mainly because of poor penetration of the spray to the growing weeds or the soil surface. Even with excellent straw management, the stubble and mulch produced interfere with the placement of the herbicide onto the soil so that weeds escape and grow. The advantage of burning the straw to improve herbicide penetration is obvious. Fortunately, the availability of postemergence herbicides is increasing and these can be employed if necessary.

Apart from locally important, hard-to-kill perennial weeds, the most serious weed problem is that presented by the volunteer growth of the small grain crop just harvested. The benefit from minimizing grain harvest loss is doubly important in a double cropping system.

If soil moisture levels are adequate, more conventional tillage methods can be used, usually a disk-harrow. Under these conditions normal herbicide practices may be employed and weed control may be as effective as in standard soybean production.

So long as maturity is not a problem, full-season cultivars can be used in a double crop system. Full-season cultivars, because of the photoperiodic response, will be somewhat shorter than when planted at normal dates and will mature only slightly later. When the planting date is delayed, a cultivar one Maturity Group earlier would give added stability for maturity but would yield less. The use of very early lines would result in short plants with low yields. In order to compensate for the reduction in plant stature inherent in planting so late, production in narrow rows is essential if satisfactory yields are to be achieved. The earlier the crop can be planted after the small grain crop is removed, the larger the plants will be and the less critical it is to go to narrow rows. Certainly the later the crop is planted, the earlier the Maturity Group of the cultivar, and the further north the crop is planted, the more important it is to narrow the row width, with the extreme being ultranarrow rows. The constraints of the planting equipment available may dictate the row width possible. Conventional no-tillage equipment, without considerable modification, does not lend itself to row widths narrower than 35–50 cm. For narrower rows, grain drills have been used; however, seed distribution within the row is not as good and, except under ideal soil moisture conditions, seed is not covered adequately to ensure even germination. Some of the newer grain drills with double disk openers, depth control devices, and individual press wheels may prove to be useful in offsetting the problems associated with the use of a grain drill for double crop purposes.

For early double crop planting the general recommendation has been to maintain the same seeding rate as for normal planting dates. However, as the seeding date is delayed and if earlier cultivars are grown on narrower row widths, there is increasing evidence that the seeding rate should be increased 15–20%. This is particularly important in the northern areas of double crop production.

The northern limit for double cropping soybeans after small grains has not yet been established. It may yet be possible to extend the practice into the central or even northern corn belt. Preliminary work indicates that some of the very early soybeans (Group 00) may be used in these areas. The fact that these types are insensitive, or nearly so, to photoperiod could permit them to grow to near-normal stature even when planted late. If small grain breeders can produce earlier maturing cultivars specifically for use in a double cropping system and if farmers are prepared to either swath or harvest their crop at higher moisture levels, the area of double cropping can certainly expand northward.

At the current northern margin for double cropping, attempts are being made to establish soybeans in standing grain (relay cropping), either by seeding soybeans by air after heading of the grain crop, or by drilling soybeans into the growing grain earlier in the season. These techniques have produced highly variable stands and yields are not considered as reliable as those from double cropping. The extreme in relay cropping would be mixed cropping. The nearest approach to this concept has been the establishment of spring grains early, followed by the interplanting of soybeans when the spring grain is 25–30 cm high. Preliminary experiments utilizing this system have been encouraging.

G. Row Width and Seeding Rate

Traditionally soybeans have been grown on almost identical systems in the northern and southern areas of production. Although cultivars differed, row widths and seeding rates were similar. This might be called the "100 cm row, 33 seeds per meter of row," formerly called the "40 inch row, 10–12 seeds per foot of row" system. Although the system appears to be appropriate for the south, its adoption in the north appears to be more a result of the legacy of the horse-powered row production system of the turn-of-the-century corn crop, from which the row width was retained after tractor power replaced the horse, than of research findings, which consistently show a yield response to narrower row widths. Corn was, and still is, king in the midwest. The management of secondary crops, like soybeans, was dictated by the corn crop. Despite the fact that soybeans would have responded to 75-cm or even 50-cm rows, with interrow cultivation, not until corn rows began to narrow, did significant areas of soybeans go onto narrower rows. Corn yields are less responsive to row widths than soybean yields.

Soybeans, like all crops, convert sunlight energy into chemical energy (dry matter). If efficient conversion is to occur, it is important to intercept as much of the light as possible. In the early part of the season, when the plants are small, much of the light is not intercepted by the leaves but strikes the soil. The critical time for high light interception occurs once the plant becomes reproductive and starts to make beans. In general, if top yields are to be achieved, the leaf canopy should close the rows as the reproductive stage is reached. Southern, determinate soybeans achieve most of their vegetative growth before the reproductive phase of growth begins. The canopy will have closed over a 100 cm row so top yields can be achieved. In the north, indeterminate cultivars are used; plants will have achieved only about half of their vegetative growth when reproduction begins. Pod production begins while the plant is still producing additional vegetation. The canopy does not close until 3–4 weeks later and yields have suffered because all of the light was not being intercepted during the reproductive period. Row widths narrow enough to close the rows at flowering time would have resulted in higher yields.

How narrow should the rows be? Several factors affect the decision, most of which are related to the degree of canopy development at the onset of reproduction.

1. *Latitude*

The further north, the shorter the growing season, the less the vegetative growth prior to flowering and the greater the need to narrow the row width. In areas of Minnesota, Wisconsin, and Ontario, Canada, where Group 00 and 0 soybeans are grown, drilled or solid-seeded soybeans are recommended for top yields. The row width is so critical that wide row production would not be profitable in most years. Even when the effectiveness of full-season chemical weed control is uncertain, the reduction in yield from wider row production would be greater than the reduction due to late-germinating weeds. The apprehension which farmers have concerning their inability to cultivate late-germinating weeds in a solid stand is, in part, offset by the fact that many late-germinating weed species cannot compete with the rapidly established, dense canopy which develops in solid-seeded soybeans. While still a future practice for most midwestern farmers, solid-seeded soybeans have been a reality for many farmers in shorter season areas for some years.

In the south the facts are clear. Conventional row widths are recommended for adapted full-season cultivars planted at the recommended time. There is little evidence to support any other advice. Only under atypical conditions would exceptions occur.

Between the extreme northern area of production and the deep south the response to row width is somewhat predictable, with increased responses occurring as one moves north. However, apprehension regarding complete dependence on chemicals for weed control and the continued use of corn planting equipment have delayed the acceptance of narrow row production on all but the innovators' fields.

2. Date of Planting

There is general acceptance, both north and south, that late-planted soybeans, because they achieve less vegetative growth and leaf area, should be planted at narrower row widths in an attempt to maintain yields.

3. Choice of Cultivar

a. **Maturity Group.** When early cultivars are planted, perhaps to achieve earlier harvest, row widths should be narrowed. Early cultivars are smaller, less leafy and hence perform better in a narrower row system.

b. **Morphological Type.** In recent years several types of "thin line" soybeans have been developed. These types have fewer branches and may have more open, erect canopies which give them the "thin" appearance. Specific recommendations for narrower row widths, when such cultivars are available, are being made in some leading soybean states. Semi-dwarf soybeans, which will soon be commercially available in the northern states, also appear to be more responsive to narrower row widths than conventional plant types (Cooper, 1976).

4. Management Level

Farmers currently producing top yields on wide rows stand to gain more by shifting to narrower rows. Obviously if good yields are being obtained already, the other management practices must be good. Row width could be the yield-limiting factor; hence a change in row width, perhaps with a change to a "thin line" cultivar, could result in a significant increase in yield. Obviously if a farmer is producing only average or below average yields, other management inputs such as increased fertility, better weed control, improved stands, or selection of a different cultivar should be tried first.

Although sounding in conflict with the above, the observation has been made that under poorer growth conditions such as insufficient nodulation, low soil moisture holding capacity or low inherent soil fertility, yields can be increased by narrowing the rows. It follows logically that vegetative growth and canopy development will be suppressed under such conditions and a response to narrower rows should be expected to occur.

5. Weed Control

The recommendation that, in those areas where row width responses occur, soybeans should be planted in as narrow a row as possible to maintain weed control, is a realistic one. However, superimposed upon this recommendation is the economic consideration of equipment use and costs. Most midwestern farmers still rely on corn planting equipment to plant soybeans, recognizing and accepting the reduction in yield in the soybean crop. As long as corn equipment is used for planting, interrow cultivation will be used to complement chemical control. Until recently, most midwestern farmers recognized the lack of stability in the dependence upon chemicals for complete control. Some newly released herbicides and some soon to be released herbicides are changing the equation. With complete chemical control of weeds a distinct reality, a complete reassessment of traditional soybean management techniques would seem to be in the offing. Content to sacrifice the reduction of 8–12% in yield inherent in growing soybeans in 100 cm rows, as opposed to 75 cm rows, it is doubtful whether farmers will continue to maintain the use of corn planting equipment when faced with the prospect of losing the 25–35% yield increase obtainable by utilizing narrow drills of 17.5–35 cm or solid seeding. The soybean would appear, at last, to be nearing release from the yield limitations imposed on it, in part, by its association with corn. Increasing numbers of farmers are currently selecting their cleanest fields, those fields where the difficult-to-kill perennials have been brought under control, for narrow row assessment. This shift can only accelerate as more reliable chemicals and newer cultivars adapted to narrow rows become available.

6. Slope of Land

There is increasing evidence that narrow row soybeans could result in less soil and water runoff than either corn or wide row soybeans. This would be particularly true if narrow rows were used in conjunction with contour planting and conservation or minimum tillage.

Seeding rate recommendations cannot be made separately from other production considerations, most particularly row width. Although seeding rate recommendations are frequently made on a weight per area basis, in actual fact these rates are usually derived from "seeds per unit length of row" research. More and more recommendations are being made using "seeds per unit length of row" (Table I).

Higher rates may be recommended when planting late, where crusting is anticipated, and/or a rotary hoe will be used. Higher populations improve the likelihood that germinating seedlings will break the crust and emerge. A rotary hoe can be used to break the crust to improve emergence. Whether rotary hoed or not, where crusting conditions are anticipated, increased seeding rates add stability where germination rate is known to be lower than 90% or when planting early in cooler soils. Germination will be slower, exposing the seeding to increased risk of insect and disease damage. Also less

6. Management and Production

TABLE I

Generally Recommended Planting Rates for Commonly Used Row Widths[a]

		Planting rate		
cm	(inches)	Seed per meter of row	Seeds per foot	Quantity (kg/ha or lb/acre)[b]
100	(40)	33–40	10–12	45–50
75	(30)	26–33	8–10	50–55
50	(20)	20–26	6–8	55–65
25	(10)	13–17	4–5	70–80
17[c]	(7)	8–10	2½–3	85–110

[a] When no-till planting or when a rotary hoe is normally used, planting rate should be increased by 10%. Planting rate should be increased to compensate for poor seed germination. For example, the planting rate should be multiplied by 100/85 for seed of 85% germination.

[b] Planting rates per unit area are approximate. Most commercial cultivars range in seed size from 5000–8000 seeds/kg. These weights of seeds per unit area assume a seed size of 6600 seeds/kg (3000 seeds/lb).

[c] Recommended in northern latitudes, for late plantings, or for thin line cultivars.

vigorous seedlings may not emerge. Higher planting rates also improve performance of "thin line" cultivars.

It is fortunate indeed that over a fairly wide range of seeding rates, yield is relatively unaffected. This adds much stability to the crop when the planter or the planter operator misfunctions and when, for one of a variety of reasons, such as poor planting, crusting, or low germination, stand is less than desired. As long as the gaps between seedlings are not too great, the soybean plants remaining have a phenomenal ability to compensate by greatly increased branching.

In general, problems arise more often with excess stand than with insufficient stand. Both extremes can result in reduced yields. Plants in thin stands tend to pod lower to the ground and produce larger side branches. At harvest, low pods and, frequently, broken lower branches are missed by the cutterbar and harvest losses are unduly high. Plants growing in thick stands tend to grow tall with thin stems. Although lowest pod height may be higher, such plants are predisposed to lodging. Early lodging reduces yield potential and increases harvest loss; late lodging increases harvest loss.

High rates of seeding result in (1) an increased tendency toward lodging, particularly in narrow rows, where natural thinning is normally less; (2) reduced branching; (3) a lower number of pods per plant; (4) a reduced number of seeds per plant; (5) a slight increase in plant height; and (6) little, if any, change in seed size, percentage protein or percentage oil.

The size of seed planted is probably not as important as the uniformity of size. Because the seed cotyledons emerge to become the first (cotyledonary) leaves, uniformity of seed size guarantees that the seedlings begin with similar photosynthetic areas. This, coupled with the even distribution of seed within the row, possible if seed is of common size, results in a more even growth and more even competition between adjacent plants. Under such conditions the attrition of seedlings by natural thinning is reduced.

Although most farmers prefer medium to large seed for planting, there is some evidence that, under heavier soil conditions, small to medium size seeds encounter less of a problem in emergence than do seedlings from large seed (Edwards and Hartwig, 1971; Burris et al., 1973).

Planting depth should be 2.5–5.0 cm. Very occasionally deeper planting (7.5 cm) can be justified on lighter soils in order to place the seed in contact with moist soil. However, even in these soils there is a risk in deeper planting. Under good moisture conditions planting should be 2.5–3.75 cm deep. Attempts to plant shallower than this may result in lack of cover of some seed and increased risk of herbicide injury. Early planted beans should not be planted deep, only deep enough to guarantee coverage. Soil temperatures may be cooler at early dates, retarding germination and increasing risk of insect and/or disease injury. Also, cultivars differ in their ability to emerge from lower depths, particularly those with short hypocotyls (Grabe and Metzer, 1969).

Uniformity in depth of planting is gaining increased importance. Seeds planted at the same depth tend to emerge more evenly and develop into plants of more uniform size if seed is evenly spaced within the row. Under such conditions competition between adjacent plants will be more even and there will be reduced attrition of the stand due to interplant competition.

Planting depth and emergence problems are not as crucial with corn as with soybeans. As a result, depth control devices on corn planters, while adequate for corn, have been marginally adequate for soybeans. As specific soybean planters are developed, adequate depth control devices must be an essential part of their design.

H. Planting Equipment

Soybeans have traditionally been planted with existing row crop equipment purchased for use with other crops on the farm. The most commonly used equipment is the unit corn planter, but in certain areas, where other cash crops are produced, cotton planters, bean and beet planters, or peanut planters are sometimes used. Occasionally grain drills are employed for soybean planting.

A good soybean planter should provide depth control within 1 cm. Systems of depth control include: (1) press wheel gauging in which the press wheel position is adjusted relative to the carrying frame; (2) gauge shoes with a turned up edge which can be adjusted in relation to the furrow opener; (3) depthbands which can be bolted to the double disk; and (4) gauge wheels which operate similarly to press wheels except they do not run over the planted row but may be placed forward of the opener or as pairs on each side of the opener. More sophisticated equipment, including sensors or ultrasonic devices to measure disk penetration and relay a signal to the hydraulic system, which instantaneously adjusts the cut to the desired depth, are in the experimental stage.

The planter also should provide a good seed spacing. Both the standard plate-type corn planter and the newer plateless planters are adequate in this respect. Where plates are used, proper plates and sized seed can greatly improve the distribution. Plate cells too large may result in many "doubles;" cells too small result in skips and/or damaged seed.

The planter should ensure good seed–soil contact. The double disk opener is probably the best device in placing seed properly. A press wheel, of the proper type and proper pressure, should firm the soil around the seed and may reduce the potential for crusting.

Most of the older grain drills are weak in all of the above requirements. Certainly seed metering leaves much to be desired, as do depth control and firming. In anticipation of the rapid shift to narrow row or drilled soybean systems arising from improved herbicide technology, several companies have experimental drill planters designed specifically for soybean planting. These new drills are large, sophisticated and expensive, as they include depth control devices, accurate metering devices, and individual press wheels. Grain drill-type planters designed for soybeans are shown in Fig. 2.

Unit planters can be used down to row widths as narrow as 37.5–50 cm. Row widths narrower than that would require a double tool bar with staggered planter units—a very costly, unwieldy piece of equipment.

Chemical weed control has created the opportunity to utilize the narrow row advantage in soybeans. Although it has been a long time coming it would appear that the truly unique soybean planter is on the threshold of reality. The introduction and mass acceptance of such equipment, more than anything else, will signal the fact that soybeans have truly achieved first-class status.

I. Inoculation

Well-nodulated plants are essential for high soybean yields. Inoculation of seed with good strains of rhizobia is the most reliable means of guaranteeing that proper nodulation occurs.

Fig. 2. Grain drill-type planters designed for soybeans. A. Press drill with center-mounted wheels pulled behind a packer-mulcher. B. Close-up showing large-diameter press wheels and depth bands on the outside of the closest two double disk openers. C. End-wheel drill following

Particularly in northern areas, it is difficult to obtain adequate nodulation the first time the crop is grown in a field. Under such situations, seed box treatments seldom give as good results as a slurry treatment which sticks many more bacteria onto the seed. Because the rhizobia will survive in the soil for several years, there is some uncertainty whether inoculation is always necessary. In general, the best nodulation occurs on fields which have grown well-nodulated soybeans in the previous year or two. For fields not in soybean production for 3–4 years prior to planting, inoculation is regarded as

6. Management and Production

a packer-mulcher. Half of the press wheels are visible. D. Close-up showing the spring-loaded press wheels and the depth bands on one side of the double disk openers. Photographs courtesy of International Harvester Corp.

cheap insurance. Fields which have not grown soybeans for 5 or more years, should always be inoculated.

Inoculation techniques vary. Commercial formulations of inoculant include those listed below.

1. Peat-Based inoculant (the most common type). This type may be applied directly to the seed in the seed box or preferably mixed onto surface-moistened seed to obtain a better seed coverage and retention of

higher numbers of bacteria. Seed should be dry when planted in order to avoid damaging the seed coat.

2. Liquid formulation. This is a liquid suspension of rhizobial cells applied into the furrow with the seed, thereby avoiding any mixing of treated seed with the inoculant.

3. Granular formulations. This type is banded via the insecticide applicator into the seed zone at planting time. Granular inoculant and liquid suspensions have the potential of greatly increasing the rhizobial numbers in the seed vicinity so may have particular value on new fields or fields known to have had nodulation problems. This formulation also avoids mixing the inoculant and seed dressings on the seed.

4. Preinoculated seed. This treatment may include only rhizobium or may also include small amounts of mineral nutrients, especially molybdenum, and/or pesticides. Preinoculation treatments should not be made too long prior to planting.

Standard peat inoculants should be applied just prior to planting. The bacteria lose their viability very quickly if exposed to sunlight, high temperature, or drying conditions. With this in mind, care should be taken to ensure that any formulation of inoculant purchased is fresh and has been stored under cool conditions. On-the-farm storage and handling should avoid exposing inoculant to high temperatures.

It is often difficult, where soybeans are grown with regularity, to show a yield response from inoculation. Soils in such situations normally have a high population of adapted strains. These strains may not be as efficient in nitrogen fixation as those contained in the inoculant, but the fact that they are present in such large numbers may result in their forming most of the nodules. Except where unusually large amounts of inoculant are applied, there is little likelihood of appreciably changing the proportion of nodules produced by the introduced strain or strains. Granular and liquid formulations represent the most likely means of altering rhizobia populations in the soil. However, because of their higher rates of application and higher cost, particularly on narrow rows, the use of granular formulations is not yet widespread. In soils of low pH, rhizobial numbers decline rapidly. Ideally this soil condition should be corrected by liming; however, where lime is not applied, inoculation should be practiced, preferably with the addition of molybdenum.

Attempts continue to isolate more efficient strains of rhizobium. It is unlikely that a new "super strain" will be found that is equally effective on all cultivars, or is it probable that such a strain would be ecologically adapted to survive over all soybean soil conditions. However, the overall economic importance of the nodulation–nitrogen fixation phenomenon, particularly in light of the predictable increase in cost of petroleum-based nitrogen fertilizers, emphasizes the importance of research directed towards improving this vital biological process.

J. Seed Dressings

Seed dressings with insecticides and/or fungicides are not generally recommended. The more northern areas of soybean production usually include seed treatment as part of the overall recommendation. In other areas seed treatment is recommended (1) where seed germination is low (85% or less), (2) when seeds are weathered or seed coats broken, (3) when planting earlier than normal, and (4) when planting into cool, moist soils which are not conducive to rapid germination and emergence. Although stand increases from seed dressings are frequently observed under the above conditions, yield responses are seldom realized. Seed treatment may be applied several weeks or months in advance of planting, but inoculant should not be added to treated seed until immediately before planting. Seed dressings, particularly fungicides, kill rhizobia if the two are in contact for several days. In new soybean fields, inoculation is more important than seeed treatment, so the latter operation is usually omitted.

Seed dressings usually consist of a mixture of insecticides and a fungicide. The fungal diseases of concern at the seedling stage include those caused by *Pythium*, *Fusarium*, and *Rhizoctonia* species. Most commonly recommended fungicides are thiram, captan, maneb, oxythiin plus thiram, chloronil, and terrachlor. Seedling insects of concern include seed maggot and, occasionally, wireworm. Diazinon and lindane are the most frequently used insecticides although several promising new insecticides are currently under test.

Two precautions should always be observed when employing seed dressings; (1) always read the label on the container and follow directions carefully and (2) never feed treated seed to livestock.

The best protection against seedling diseases and insects is the use of high quality, vigorous seed.

III. MANAGEMENT DURING THE GROWING SEASON

A. Selection and Use of Herbicides

For many growers, inadequate weed control is a major limiation to high yields. Weeds are estimated to reduce the United States soybean crop by 17%.

TABLE II

Examples of Various Classes of Herbicides

Class	Chemicals
Preplant incorporated	Trifluralin, nitralin, vernolate, dinitramine, profluralin, trifluralin + metribuzin
Preemergence	Alachlor, linuron, chlorbromuron, oryzalin, naptalam + dinoseb, chloramben, metribuzin
Seedling emergence	Dinoseb, naptalam + dinoseb
Early postemergence, broadcast	Bentazon, chloroxuron
Directed postemergence	Dinoseb, linuron, paraquat, 2,4-DB
Late postemergence	2,4-DB

Chemicals for weed control started to be used in the 1950's. Commercial use of herbicides expanded during the 1960's and in the 1970's the effectiveness and use of herbicides has continued to increase. Herbicides can be divided into four broad categories: those applied as preplant incorporated treatments, as preemergent applications, at seedling emergence or "cracking" sprays, or as postemergent applications. Examples of each class are shown in Table II. Herbicides that are incorporated are generally less dependent on moisture after application to be effective than are compounds sprayed on the soil surface. However, incorporation requires extra energy consumption, although many operators mount the sprayer nozzles on the front of a disk and incorporate the herbicides during tillage operations. Nutsedge (*Cyperus* spp.), one of the world's worst weeds, is best controlled by incorporated herbicides.

Preemergent applications are sprayed on after planting but before the crop has emerged, whereas "cracking" stage chemicals are applied just as the soybeans are emerging. Some of the chemicals used as postemergence sprays are toxic to soybeans and must be applied in a spray directed so that the soybeans are not sprayed or are sprayed only near the base of the plant. Bentazon (Basagran) has come into widespread use as an overall postemergent spray. It is used to control cocklebur and other broadleaf weeds, but gives little control of grasses. A number of effective postemergent grass herbicides are being tested. When these become available, soybean growers will have an option of waiting to see how severe their weed problems are, before they spray.

Many of these herbicides are used most effectively in combinations. Some are effective on certain weeds while others will kill weed species missed by their partner. For example, trifluralin alone gives poor control of cocklebur and ragweed plus some other weeds, but a tank mix of trifluralin plus metribuzin improves control of cocklebur and ragweed, and sometimes causes less burning of soybeans than metribuzin applied alone as a preemergent spray.

Special programs and herbicide combinations may be necessary to control the most serious weed problems. These include cocklebur, morning glory, sicklepod, nutsedges, and perennial weeds.

Soybean herbicides must be applied within a fairly narrow range of rates. If rates are too low, weed control will be inadequate. Application of herbicides at rates which are too high can damage or kill the seedlings. In addition, disking too deep when incorporating pre-plant incorporated herbicides can increase plant damage (Kust and Struckmeyer, 1971), as can planting seed at too great a depth. Herbicide manufacturers' directions on the container label must be strictly followed because of the fairly narrow range of suitable herbicide rates.

Soybean cultivars differ in their tolerance to some herbicides. The main cause for concern is burning injury with the herbicide metribuzin, sold commercially as Sencor and Lexone. Certain cultivars including TRACY, COKER 102, SEMMES, ALTONA, and VANSOY are particularly susceptible (Andersen, 1976).

Residues from applications of herbicides to previous crops can cause soybean production problems. Atrazine applied to a previous corn crop can reduce soybean stands and yields. Small-seeded cultivars are most susceptible. Soybeans can tolerate more atrazine residue than can cereals. Some reports indicate that atrazine levels up to 1.12 kg/ha in the soil did not greatly reduce yields (Andersen, 1976), but levels above 0.2 kg/ha can be expected to cause some damage. In addition, soybean herbicides may interact with the atrazine to cause damage. For example, Anderson (1974) applied 0.37 kg/ha of atrazine to six cultivars of soybeans, as a preemergence treatment. None was severely affected. However, when 1.12 kg/ha of linuron was also applied, the small-seeded cultivar VANSOY suffered a severe loss in stand. If linuron–atrazine combinations increase soybean damage, then soybean herbicides with similar structures, including metobromuron, chlorbromuron, and metribuzin, also might be expected to increase damage on soils with triazine herbicide residues (G. W. Anderson, personal communication).

Details of the use of herbicides are well reviewed by Wax (1973). In addition, soybean states have advisory bulletins giving up-to-date information on herbicides and their use.

B. Lodging

In highly productive environments soybean plants usually produce rank vegetative growth so that lodging occurs. Lodging refers to the bending of the main stem so that the plant falls or leans over. Conditions favoring vegetative growth promote lodging. These conditions include high available soil nitrogen, abundant water, and warm weather. High populations and narrow rows also contribute to lodging because the increased competition at high populations results in reduced stem diameters and spindlier plants. Lodging also is increased by heavy rains and severe winds.

When lodging occurs in the early stages of pod filling, yield reductions of from 13 to 32% occur (Woods and Swearingen, 1977). Yield reductions are mainly due to decreased numbers of pods per plant, rather than smaller seed or fewer seeds per pod. When the plant lodges it loses apical dominance and growth of branches is enhanced. This vegetative growth probably increases flower and pod abortion because less photosynthate is available to reproductive structures. Other reasons for yield reductions due to lodging include poorer light distribution within the leaf canopy and poorer air circulation in the canopy, decreasing CO_2 supply and contributing to higher incidence of disease. Another major reason for lower yields in lodged soybeans is that greater combine losses occur at harvest time because part of the crop goes underneath the cutterbar. When lodging occurs late in the grain-filling period, harvest losses seem to account for most of the yield decrease.

Some management decisions can reduce lodging. In situations where lodging is expected, lowering the seeding rate increases stem diameter and helps plants stand better. Wider row widths also reduce lodging (Cooper, 1977). However wide rows also decrease yields. Some cultivars are less susceptible to lodging than others. Several breeding programs are attempting to breed semi-dwarf characteristics into indeterminate soybeans to reduce lodging in productive environments (e.g., Cooper, 1976). Some growth regulators can shorten plants and reduce lodging. The chemical triiodobenzoic acid (TIBA) was used for this purpose (Greer and Anderson, 1965), but its use has been discontinued. However, research workers continue to search for growth regulators and genotypes that will allow the soybean crop to utilize highly productive environments more fully.

C. Diseases

As the area planted to soybeans has expanded, diseases have increased in number and severity. Over 100 diseases are known to affect soybeans and about 35 are of economic importance. Chemicals for soybean disease control are as yet rarely used.

Soybean diseases will not be discussed in great detail here. A recent compendium of soybean diseases (Sinclair and Shurtleff, 1975) provides excellent coverage of diseases and nematodes (See also Chapter 4, Section VI,

B). Diseases of soybeans can be classified by the plant part they affect and whether they are caused by fungi, bacteria, or by a virus. The most serious diseases, along with their symptoms and control, are presented in Table III. Severity of the diseases will depend on host plant resistance to the pathogen and also on the environmental conditions. Most fungal and bacterial diseases of aerial parts are more severe under warm, moist, humid conditions when growth and multiplication of the pathogens are favored.

D. Nematodes

Plant parasitic nematodes are microscopic animals. They inhabit the soil as eggs and at early larval stages. Adults are usually eel-shaped but some may become swollen in later stages. Nematodes that feed on plants have an oral spear for feeding, which punctures plant cells. In some cases, such as with root-knot and cyst nematodes, the host plant responds to the injury by producing enlarged cells where nematodes feed. With root-knot nematodes, other cells are produced as well, producing galls on the root. Other nematodes, such as the sting types, feed on cells near the root tip and inhibit root extension.

1. Root-Knot Nematode

Root-knot is caused by several species of *Meloidogyne* and has been reported from most soybean-growing areas in the world. Yield losses from 30 to 90% have been reported for susceptible cultivars (Sinclair and Shurtleff, 1975). Losses are more severe in warm climates and coarse-textured soils. Typical symptoms are stunting, yellowing, wilting under moisture stress, and the development of spherical or spindle-shaped galls on the roots. *Meloidogyne incognita*, the common southern root-knot nematode, causes large galls up to 20 mm in diameter. The northern root-knot nematode, *M. hapla*, which occurs frequently in northern and mid-Atlantic states, causes small galls (Good, 1973). *Meloidogyne javanica*, the Javanese root-knot nematode, and *M. arenaria*, the peanut root-knot nematode, also damage soybeans in the southern United States.

Root-knot nematodes have broad host ranges, including many weeds and commercial crops. Several root-knot nematode species can infect soybeans at the same time. If only one species is present and a nonhost crop is grown in long rotations with soybeans, some control can be obtained. Cultivars differ in their susceptibilities to root-knot nematodes. Proper cultivar selection can minimize losses. Soil fumigation with nematicides can be effective but fumigation is advisable only if tolerant cultivars are unavailable and production returns make it economical.

TABLE III
Descriptions of Soybean Diseases[a]

Type of pathogen	Plant part affected	Common name	Pathogen	Symptoms	Damage estimate	Location	Control
Fungus	Root and stem	Phytophthora rot	*Phytophthora megasperna* var. *sojae* (9 races identified)	Wilting and death of young plants. Discolored stems and roots, rotting roots, leaves yellow and wilt in older plants.	Complete kill in some areas of field.	Poorly drained, fine-textured soils from Ontario to Mississippi.	Resistant and tolerant varieties, good drainage.
		Rhizoctonia	*Rhizoctonia solani*	Reddish-brown decay of outer cortex of stem base and older roots. Plants wilt and die if severe.	Stand losses of up to 50% and yield losses of up to 40% reported.	Poorly drained areas; fine-textured soils, worldwide.	Fungicides improve emergence. Good drainage. No resistant varieties but differing susceptibilities.
		Stem canker	*Diaporthe phaseolorum* var. *caulivora*	Reddish-brown, lesions at leaf scars of lower nodes expand to brown or black cankers that girdle the stem.	Yield losses of 20–50% can occur if infected at early pod stage.	Prevalent in the north-central soybean growing area in North America.	Crop rotation. High quality seed. Less susceptible varieties (none are resistant).

Brown stem rot	*Phialophora gregata*	Dark reddish-brown vascular bundles and pith when stems are split; premature leaf death.	Yield losses of 25% or more.	Greatest damage occurs with a cool period during pod filling followed by hot, dry weather. Reported in North America and several other countries.	Crop rotation 4 years long. Resistant varieties are being bred.
Pythium root rot	*Pythium* species	Seed rot, seedling root rots, damping off; apical growth stunted.	Slight economic loss.	Scattered plants; worldwide, with broad host range.	High quality seed. Seed protection with fungicides. Plant in warm soil.
Fusarium root rot	*Fusarium* species	Rotting of seedling roots, usually in outer tissue.	Stands can be reduced by 2/3. Yield reductions of 50% reported.	Develops in cool weather (10–15°C) on young plants. Worse in poorly drained soil.	Resistant varieties. Use high-quality seed in warm soil with good drainage.
Charcoal rot	*Macrophomina phaseolina*	Black streaks on woody portion of roots and stems. Split roots and lower stem show small, black sclerotia.	Stand reductions can occur, but damage usually is not severe on vigorous plants. Disease suppresses yield.	Normally occurs after midseason, in warmer areas, after unfavorable weather. Worldwide with many hosts.	Maintain good growing conditions, low populations, and long rotations.

(continued)

TABLE III

Descriptions of Soybean Diseases[a] (continued)

Type of pathogen	Plant part affected	Common name	Pathogen	Symptoms	Damage estimate	Location	Control
		Sclerotia blight	*Sclerotium rolfsii*	Cottony mycelial growth on stems with brownish sclerotia on surface. Yellow leaf spots may occur. Usually occurs in patches.	Does not cause major losses except in tropics.	Occurs mainly on lighter soils in warm areas. Worldwide.	Most varieties have some tolerance or resistance. Crop rotation. Deep plowing.
		Sclerotinia stem rot	*Sclerotinia sclerotiorum*	Older plants wither and die with leaves still attached. Cottony mycelial growth occurs on stems, branches and pods. Large sclerotia form on the stem and in stem pith.	Usually of minor importance. Under proper conditions fields show damaged patches. Seed containing sclerotia often are rejected for export.	Reported from many areas in the world. Occurs under extended moist, humid conditions.	Plant well-cleaned seed. Avoid solid seedings and tall, lodging cultivars. Keep canopy open.

Fungus	Root and stem	Anthracnose	*Colletotrichum dematium* and *Glomerella glycines*	Damping off of infected seedlings. Older stems and pods may show black fruiting bodies. Pods and lower branches die in serious infections.	Serious damage may occur under disease-promoting conditions. Reductions occur in stand, yield and quality. Damage usually slight in United States.	Observed in all major soybean areas. In United States extends from southeast to midwest. Serious in warmer areas, particularly during rainy periods.	Sow disease-free seed. Treat seed with fungicide. Plow under crop residue.
Fungus	Leaf	Brown spot	*Septoria glycines*	Angular reddish-brown leaf spots occur, followed by premature leaf yellowing and drop. Spots progress up the plant.	Causes premature defoliation and some loss in yield.	Quite prevalent in cooler areas of production. Worldwide. Most prevalent in continuous soybeans and under moist conditions.	Crop rotation. Plow under residues. Disease-free seed.
		Frogeye leafspot (Cereospcra leafspot)	*Cercospora sojina*	Angular reddish-brown lesions with gray centers and no yellow border. Elongated lesions may occur on stems and pods.	Up to 15% on susceptible varieties.	Worldwide. Usually occurs in warmer areas of United States, especially during warm, humid weather.	Resistant varieties. Disease-free seed. Crop rotation.

(continued)

TABLE III

Descriptions of Soybean Diseases[a] *(continued)*

Type of pathogen	Plant part affected	Common name	Pathogen	Symptoms	Damage estimate	Location	Control
		Target spot	*Corynespora cassiicola*	Target-like lesions, irregular reddish brown, surrounded by dull green or yellowish-green halo. Severely infected leaves drop prematurely.	Susceptible varieties in southern United States reported to lose 18–32% of yield.	Occurs in southern United States and other warm areas. Develops most with high moisture during pod filling.	Most varieties grown in the southeastern United States are tolerant.
		Powdery mildew	*Microsphaera diffusa*	Ash-gray spots on upper leaf surface soon become covered with a powdery mass. Underlying tissue turns pink to red.	Occasionally severe if it occurs early during the season.	Widespread but a problem only under humid, moderate temperature conditions.	Resistant varieties. Foliar fungicides.
Fungus	Leaf	Downy mildew	*Peronospora trifoliorum*	Yellow to brown irregular lesions on upper leaf surface. Gray tufts on lesions on lower leaf surface, especially in	Reduces seed quality and can reduce yield slightly.	Worldwide. Develops in moist, warm weather.	Plow under residues. Crop rotation. Disease-free seed. Some resistant varieties. Seed treatment helps.

				moist weather. Dull white crust on seed surface.		
		Phyllosticta leafspot	*Phyllosticta sojaecola*	Round to oval leafspots or V-shaped lesions from leaf margins, dull gray to tan.	Minor importance recently.	Worldwide. Mainly on young plants.	Disease-free seed. Crop rotation. Plow under residues
		Rust	*Phakopsora pachyrhizi*	Chlorotic to gray-brown spots. Pustules break open and release rust-coloured spores. Premature defoliation and reduced pod and seed development.	Severe losses where prevalent.	Widely distributed, especially in Eastern Hemisphere. Not prevalent in United States.	None satisfactory. Fungicides may help. Resistance being sought. Poses threat for United States production.
Fungus	Seed	Purple stain	*Cercospora kikuchii*	Purple seed discoloration. Infected leaves develop angular reddish-brown spots late in the season.	Reduces grade but not yield. May reduce germination.	Worldwide. More severe with warm, moist conditions during seed development.	Moderately resistant varieties. Disease-free seed. Fungicides for seed and foliar application.

(*continued*)

TABLE III
Descriptions of Soybean Diseases[a] (continued)

Type of pathogen	Plant part affected	Common name	Pathogen	Symptoms	Damage estimate	Location	Control
		Pod and stem blight	*Diaporthe phaseolorum* var. *sojae*	Seed is shriveled and may have white mycelium on surface. Reduced seed germination. Small black specks, often in rows, on lower stems, branches, and pods as plants reach maturity.	Reduces grade but not yield.	Worldwide and common. Increased by wet weather at harvest time.	Disease-free seed. Crop rotation. Plow under residues. Fungicides for infected seed. Varieties differ in resistance.
Bacteria	Leaf	Bacterial blight	*Pseudomonas glycinea*	Brown or black dead area surrounded by angular, water-soaked area and yellow-green halo. Stems, pods and seeds can be infected. Leaves have a tattered appearance.	Most serious bacterial disease. Reports of essentially total crop loss come from Russia.	Worldwide and very common. Most prevalent in early season and cooler regions; decreased by hot, dry weather and increased by wet conditions.	Varieties differ in susceptibility. Disease-free seed, crop rotation. Plow under residue. Avoid cultivation when foliage is wet.

Bacterial pustule	*Xanthomonas phaseoli* var. *sojensis*	Pale green spots with elevated red-brown centers, in which a raised, light-colored pustule usually forms. Spots yellow and coalesce to form brown, dead patches but there are no water-soaked areas. Leaves have a ragged appearance.	Premature defoliation reduces yield. Severity depends on conditions.	Worldwide and very common. Most prevalent in southern United States areas and with frequent rainfall.	Resistant varieties. Disease-free seed. Crop rotation. Plow under residue. Avoid cultivation when foliage is wet.
Wildfire	*Pseudomonas tabaci*	Brown spots of varying shapes and sizes surrounded by a broad yellow halo. In damp weather, dead areas enlarge and tear away, producing tattered leaves.	Presently considered of minor importance.	Disease is known worldwide on tobacco but is reported only from United States and Brazil for soybeans. Often occurs as a secondary infection after bacterial pustule or blight.	Varieties resistant to bacterial blight and pustule are also resistant to wildfire and control methods are similar.

(continued)

TABLE III

Descriptions of Soybean Diseases^a *(continued)*

Type of pathogen	Plant part affected	Common name	Pathogen	Symptoms	Damage estimate	Location	Control
		Bacterial wilt	*Pseudomonas solanacearum* and a *Corynebacterium* sp.	Wilting of leaves, particularly with high moisture demand.	Minor importance.	*P. solanacearum* wilt was reported from North Carolina and Russia. *Corynebacterium* wilt was reported from Iowa.	Sanitation as for bacterial blight.
Virus	Leaf	Soybean	Soybean mosaic virus (SMV)	Stunted plants with misshapen, puckered leaves which curl downwards at the margins. Mottled seeds and curled, thin pods.	Yields may be reduced 25% or more in the field. Seed quality is reduced.	Worldwide, seed-borne, aphid-transmitted, and with a wide legume host range. Symptoms are most expressed at temperatures below 20°C.	Sow virus-free seed. Control weeds and aphids. Separate soybeans from other legumes.
		Bean pod mottle	Bean pod mottle virus (BPMV)	A mild chlorotic mottle occurs on young leaves and is masked as leaves get older.	Yield reductions may be 10–15% from BPMV, but up to 60%, if plants also are infected with SMV.	First reported in United States in 1958. Seed-borne and beetle-transmitted. Symptoms are most obvious during rapid growth and cool conditions.	Same as for SMV. Spraying for leaf-eating beetles and early planting may help.

Bud blight	Tobacco ringspot virus	Curving of the young plant's terminal bud to form a crook, brown pith, stunted plants, dwarfed, cupped or rolled leaves, and dark blotches on pods.	Yield losses of 25–100% may occur, depending on time and severity of infection.	Found in United States, Australia, Canada and China. Seed-borne with no known efficient insect vectors. Spreads into soybeans from a wide host range.	Grow a tall nonhost crop such as corn around crop. Kill weeds around crop before emergence. Sow virus-free seed.
Yellow mosaic	Bean yellow mosaic virus (BYMV)	Young leaves show yellow mottling, slight puckering and sometimes a yellow band along major veins developing into rusty necrotic spots.	Yield reductions are not known.	Reported from many areas in United States and from Japan. Widely distributed in United States midwest. Insect-transmitted but not seed-transmitted. Wide legume host range.	Same as for SMV. Sinclair and Shurtleff (1975) report resistant varieties.

[a] Sources of information are Athow (1973), Dunleavy (1973), Kennedy and Tachibana (1973), and Sinclair and Shurtleff (1975).

2. Soybean Cyst Nematode

The soybean cyst nematode, *Heterodera glycines*, is a serious threat to soybean production in Japan, Korea, Manchuria, and in the United States, from the midwest to the southeast. High populations have caused losses as high as 90%. With heavy infestations, symptoms are severe stunting and yellowing of the tops. This occurs because formation of root nodules by *Rhizobium japonicum* is inhibited and the plant becomes nitrogen-deficient. Wilting and premature defoliation also can occur. Heavy infestations reduce root growth and nodules are rare. Brown cysts, slightly smaller than a pinhead, can be seen on the roots. Symptoms are most serious in sandy, dry soils with low fertility.

Four races are known and soybean cultivars have differing susceptibilities (Sinclair and Shurtleff, 1975). No resistance to race 4 is known. For other races, resistant cultivars offer control. Long rotations with nonhost crops such as cotton, corn, or sorghum are a simple control measure. Soil fumigation at planting time also is effective, but more expensive.

3. Other Nematodes

The reniform nematode, *Rotylenchulus reniformis* causes galls, stunting, and chlorosis. Reniform nematodes have caused damage in South Carolina and are considered of great potential importance in the tropics. Control is by resistant cultivars and nematicides.

Two nematodes which do not form galls are sting nematodes and lance nematodes. Sting nematodes are most common on coarse-textured soils in the southeastern and Gulf Coast states in the United States. The sting nematodes, *Belonolaimus gracilis* and *B. longicaudatus*, have caused yield reductions of more than 80% (Good, 1973). The lance nematode, *Hoplolaimus columbus*, causes severe damage on sandy soils with hardpans and is restricted to the southeastern United States. *Hoplolaimus galeatus* occurs throughout the United States soybean production area but does not cause obvious damage. Both sting and lance nematodes cause dark, shrunken lesions near the root tips and along the main root. Root growth often ceases and roots may break off, causing a stubby root appearance. Lateral roots often proliferate above the damaged areas.

Root-lesion nematodes attack the root cortex and do not form galls. If roots are severely infected, plants become yellow and stunted. Root-lesion nematodes (*Pratylenchus* spp.) can cause 25% reductions in root growth (Sinclair and Shurtleff, 1975).

E. Insects

Soybean damage by insects is most severe in warm climates and is more prevalent in the southern United States than in the north. Colder winters and shorter growing seasons in the north restrict the development of severe infestations. A wide range of insect pests are found in soybean fields but few are of economic importance. (See also Chapter 4 Section VI, C.) Soybean insect pests include leaf, stem, pod, root, and seed feeders. Leaf damage is the most obvious damage in the growing crop but frequently does not reduce yield. Turnipseed (1973) found that removal of 17% of the leaves at any stage of development did not cause yield reductions. Defoliations of 33% at midbloom did not decrease yields in any of ten tests, although later defoliations of 33% did reduce yields. Soybean plants normally produce more leaf area than that required for maximum light interception and photosynthesis, so some defoliation by insects can occur without causing yield reduction. Therefore, the grower needs to know the insect species present and how prevalent the insect is. This is best determined by shaking insects off the plants onto cloth or paper on the ground. The insects present per meter of row should be determined at several random sites across a field. Levels of insects which make spraying economical depend on the stage of development and location. Up to blooming, 33% defoliation is the threshold where treatment is suggested. After blooming, levels of tolerance drop to 15–20% defoliation.

Corn earworms and stinkbugs are damaging pod feeders. Generally, treatment is suggested when insect levels average one to three per meter of row. A brief description of the major insect pests is in Table IV. Detailed descriptions are available for the insect pests of the United States (Turnipseed, 1973; Turnipseed and Kogan, 1976) and Brazil (Panizzi et al., 1977).

F. Hail Damage

Soybeans in the United States midwest are frequently damaged by hail. The effects of hail damage at various stages of growth have been studied in several states and the results are used by crop insurance adjusters to estimate damage. Through most of the growing season, damage is estimated from the number of dead plants and the percentage of leaf area destroyed.

Soybeans can regrow after damage if the growing point at the top of the plant is intact. In addition, regrowth can occur from buds in the leaf axils, including buds at the cotyledonary nodes. If the plant is broken off below the

TABLE IV
Characteristics of Insect Pests of Soybeans[a]

Type	Common name	Scientific name and description	Symptoms and injury
Leaf-feeding beetles	Mexican bean beetle	*Epilachna varivestis*. Looks like a ladybug with yellow to copper body and up to 16 black spots on the back of an oval-shaped body. Larvae are yellow and spiny. Pupae are lemon yellow.	Larvae and adults feed on the lower surface of leaves between veins, giving leaves a lace-like appearance. Remaining veins turn brown. Most severe in eastern seaboard and in the south of the United States.
	Bean leaf beetle	*Cerotoma trifurcata*. Reddish-brown beetle, often with 4 black spots on back, and about 5 mm long. Larvae and pupae are in the soil.	Adults feed on lower leaf surface, making small holes. Damage usually occurs in August and September. Economic damage usually occurs only in the south in the United States.
	Japanese beetle	*Popillia japonica*. Adults are about 13 mm long, metallic green to bronze with black head and thorax. Larvae live in soil and look like small white grubs.	Adults are general feeders on many plant species. They eat interveinal areas in soybeans in June to August. Normally most prevalent in north-central states and mid-Atlantic states in the United States.
	Blister beetles	*Epicauta vittata* is the striped blister beetle. It has a long (2.5 cm) narrow body, long legs, striped yellow and black wings, a narrow thorax and broader head. A mashed insect can burn the skin.	Larvae feed on grasshopper eggs, so damage usually moves in from roadsides and ditches. Defoliation within a few days can occur in isolated spots.
	Cucumber beetles	*Diabrotica undecimpunctata howardii* and other species. They are all about 6 mm long, greenish-yellow, with black spots, stripes or bands.	Frequently feed on soybean leaves but rarely cause economic injury.

Leaf-feeding moth caterpillars	Green cloverworm	*Plathypena scabra.* Slender green caterpillars 3 cm long with faint white stripes along the abdomen; three pairs of ventral prolegs plus one pair of anal prolegs; slate grey to brown moth with 2.5 cm wingspan.	Larvae feed on upper foliage, causing ragged leaves and complete stripping if severe. This pest is found throughout soybean areas but is more serious in the south. A fungus pathogen often gives control. Recommended insecticides also give control.
	Soybean loopers	*Pseudoplusia includens.* Larger than cloverworms with a thick body tapering to the head. Green with lengthwise lighter stripes; two pairs of ventral prolegs and one anal pair; brown moths with a silver wing spot.	Loopers cause extensive foliage loss and some pod damage. Usually damage is economic only in the southern part of the United States. Standard insecticides are only partially effective. Fungus diseases also control populations.
	Velvetbean caterpillars	*Anticarsia gemmatalis.* Larvae are usually greenish with a prominent light stripe, up to 5 cm long; four pairs of ventral prolegs and one anal pair; light brown moths with oblique dark lines across both sets of wings.	This is a pest of southern areas. It overwinters in the tropics and migrates to temperate climates. Foliage can be stripped in just a few days. A fungus disease often controls populations and most insecticides are effective.
	Corn earworm	*Heliothis zea.* Light green to black caterpillar with light and dark lateral stripes and lighter underneath; up to 4 cm long, four pairs of ventral prolegs and one anal pair. Moths are light brown with irregular lines and dark wing tips.	Feeds on foliage but is primarily a pod feeder. Pod and foliage feeding is more prevalent in the southeastern United States. Wide range of host plants. Several broods per season. Control with insecticides is easier on soybeans than other crops because of exposed location. Fungi and predators also help control.

(*continued*)

TABLE IV

Characteristics of Insect Pests of Soybeans[a] (*continued*)

Type	Common name	Scientific name and description	Symptoms and injury
	Armyworms	*Spodoptera* species. Dark brown body with variable markings; black head; up to 5 cm long; four pairs of ventral prolegs and one anal pair. Moths are grayish-brown with white markings and have about 3 cm wingspan.	Foliage feeder, mainly in July and August. Causes webbing on leaves. Damage can be serious if it starts on small plants. Damage is more prevalent in late-planted, weedy fields. Recommended insecticides give control.
	Garden webworm	*Loxostege rantalis*. Greenish to yellowish green larvae, with black dots; about 2.5 cm long; four pairs of ventral prolegs and one anal pair. Larvae feed under a web. Moth is small and light-colored with green markings.	This is a general feeder which sometimes attacks soybeans. Heavy infestations in young plants can cause serious injury. Pigweed is an alternate host and should be removed.
	Saltmarsh caterpillar	*Estigmene acrea*. Wooly yellowish and brown caterpillar up to 5 cm long. Tiger moth adult.	Foliage feeder, more prevalent in irrigated areas. Other yellow woolybear larvae also sometimes cause economic damage.
Other leaf feeders	Grasshoppers	Several species. *Malanopus* spp. most frequent. Green to brown; large, prominent hind legs; jumping and flying ability.	Foliage feeders, particularly under dry conditions. Damage starts along fencerows, roadsides, and ditchbanks.
	Thrips	Several species. *Sericothrips variabilis* most prevalent. Minute, fragile insects about 1.3 mm long; slender with alternate brown and white bands on abdomen, four feather-like wings on adults.	Leaves are pierced near midrib and have speckled appearance. Economic damage rare.

	Leafhoppers	Potato leafhoppers (*Empoasca fabae*) are 6 mm long, green to brown, and jump when disturbed; wingless nymphs and winged adults.	Leafhoppers feed on soybean leaves, but do not cause economic damage on varieties with normal pubescence.
	Mites	Mites are tiny arthropods with four pairs of legs and not true insects. Adults are yellow to green to brown to red. Females spin a fine web on underside of leaf. Leaves have tiny chlorotic spots and a peppered appearance.	Damage is more prevalent in dry seasons and in droughty areas. Mites move in from the edges of the field. Damage can be serious if it occurs during the bean-forming period.
Pod-feeding insects	Stinkbugs	*Nezara viridula* (Southern green stinkbug), *Euschistus servus* (Brown stinkbug), *Acrosternum hilare* (Green stinkbug). Adults are shield-shaped, green or brown, 1.5 cm long and winged. Nymphs are more rounded and lack wings. Foul odor when disturbed.	Nymphs and adults pierce pods and seeds and suck plant juices. Feeding on young pods can cause shrivelled seed and pod abortion. Later damage causes small, wrinkled, misshapen and stained beans. Early, severe infestation can cause complete yield loss. Damaged seed can receive a lower grade. Insecticide control should be considered at one bug in 0.9 m of row.
	Corn earworm	*Heliothis zea* (see description under leaf-feeders)	Generally considered the most important soybean insect pest in the United States, particularly in the south. Damage is more likely to be severe adjacent to infested corn or cotton fields. Caterpillars feed on pods during seed development. Insecticide control should be considered at one caterpillar in 0.9 m of row.
Stem-feeding insects	Three-cornered alfalfa hopper	*Spissistilus festinus*. Adults are 6.4 mm long, green, triangular, blunt at the head and pointed behind. Nymphs are similar in shape but without wings.	Young stems and petioles can be girdled. Usually yield reductions do not occur unless high winds late in the season cause weakened stems to break.

(*continued*)

TABLE IV

Characteristics of Insect Pests of Soybeans[a] *(continued)*

Type	Common name	Scientific name and description	Symptoms and injury
	Lesser cornstalk borer	*Elasmopalpus lignosellus*. Larvae are yellowish-green with reddish-brown cross-markings. They tunnel into stems at the soil surface, leaving a tunnel-like webbing. Moths are brown with a 2.5 cm wingspan.	Damage to soybeans is restricted to the south in the United States. Sandy soils, dry weather, and late planting increase damage. Tunnelling kills smaller plants, resulting in poor stands. Older plants may break off.
	Cutworms	Several species. Thick-bodied larvae are up to 4 mm long, light brown to black, sluggish and roll into a ball when disturbed.	Damage occurs when cutworms eat through stems of seedlings at the soil surface. Poor stands can result, particularly after an early vegetable crop.
	Seedcorn maggot	*Hylemya platura*. Larvae are white bodied maggots which eat out interiors of seeds. Adults are small grey-brown flies.	Damage to seed can cause poor stands and yield reductions. This is a problem in northern areas where seed takes longer to emerge. Cool, wet weather can increase damage. Control is usually with diazinon seed dressing.
Root-feeding insects	Grape colaspis	*Colaspis flavida*. Larvae are 3 to 4 mm long, white, with hard yellow or brown heads. Adults are light-brown beetles about 4 mm long with ridges in wing covers.	Larvae feed on young roots, can cause yellowing and reduce stands. Damage occurs in the United States midwest and midsouth, usually in second year beans or after clover.
	White grubs	*Phyllophaga* species. Larvae are large, C-shaped grubs, white with a black head. Adults are dark brown June beetles.	White grubs eat off lateral roots. Infestation and damage occur in patches and can be severe. Under moisture stress plants wilt and die. Damage usually is worst after grass sod. Chemical controls are not usually recommended.

[a] Sources: Turnipseed, 1973; Anon, 1974; Brooks, 1976.

6. Management and Production

TABLE V

Effect of Soybean Leaf Destruction by Hail at Different Growth Stages on Final Yield[a]

Number of completely unrolled trifoliolate leaves	Percentage of leaf area destroyed			
	25	50	75	100
	Percent reduction in seed yield			
2	1	5	6	21
4	1	6	7	23
6	2	6	8	26
8	2	7	9	28

[a] From Hicks and Miller (1975) and National Crop Insurance Association.

cotyledonary node, no regrowth can occur. However, if the top of the plant is damaged or cut off above the cotyledonary node, regrowth can occur from one or more of the axillary buds. At early stages of growth, severe defoliation causes only minor reductions in yield. Table V shows yield reductions caused by defoliation at various stages of growth. Up to 75% leaf loss during vegetative growth causes less than a 10% reduction in yield in either determinate or indeterminate cultivars. Soybean yields are reduced more when complete defoliation occurs. The plant can produce new leaves and yield well enough to make replanting uneconomical, providing most plants can regrow. In a normal plant stand of forty plants per m^2, which is about equivalent to a 67 kg/ha (1 bushel/acre) seeding rate, if 20% of the plants are killed the yield reduction will be less than 10%. The soybean plant can compensate well unless there are big gaps in the stand or the damage occurs after vegetative growth has finished. The effects of stand loss and leaf loss are additive, so a 50% leaf loss plus 20% dead plants would reduce yields by about 15%. Replanting more than 30 days after the original planting date would reduce economic returns more than 15% in most midwestern United States locations. Therefore, fairly severe damage must occur before replanting should be considered.

G. Minor Elements

During the growing season producers may see symptoms of minor element deficiencies. These minor elements include manganese, iron, zinc, copper, boron, molybdenum and cobalt. Usually these elements occur in soil in quantities adequate to support good yields, but some soils may be deficient. Symptoms of deficiencies are given here. Soil application of minor elements is discussed in Section II, C.

Manganese deficiencies normally occur on sandy or gravel-based soils with pH levels near 7.0 or above. During vegetative growth leaves develop yellow to white areas between veins, which remain green. The symptoms occur because manganese is required in photosynthetic reactions and in transformations of nitrate to ammonia and protein. The deficiency is best corrected by spraying leaves with solutions of manganese sulfate or manganese oxide.

Soils with high pH and high lime levels may exhibit iron deficiency. Symptoms of iron chlorosis are stunted plants and bright yellow leaves with green veins. Iron is required in chlorophyll synthesis and in respiration. Some cultivars are more susceptible than others. As the root system expands, deficiency symptoms apparent in the young seedling may dissappear. Foliar application of a ferrous sulfate spray or iron chelates is used to correct the deficiency.

Molybdenum availability is decreased in acid soils and deficiencies may occur at pH values below 6.0. Molybdenum is required in nitrogen fixation and symptoms are the pale yellow color of nitrogen deficiency.

Zinc deficiency symptoms are light brownish-bronze leaves, particularly in cool, wet weather. Older leaves are affected most. Zinc deficiency is most frequent with heavy phosphorus fertilization.

Copper deficiency may cause severe stunting of growth, or reduce yields without obvious symptoms. Copper enhances enzyme activity and is involved in oxidation–reduction reactions. Deficiencies usually occur on muck or sandy soils.

Boron deficiency may occur in alkaline soils and on soils high in organic matter. Alfalfa is more sensitive to boron deficiency and is often used as an indicator plant. Boron deficiencies can be aggravated by applications of lime or potash. Deficiency symptoms are stunted, irregular growth because boron is necessary for normal cell division and enlargement. Applications of small amounts of boron can correct the symptoms. High levels of boron can be toxic to soybeans, particularly on soils low in calcium.

Cobalt can also cause toxicity to soybeans. Both cobalt and molybdenum can be added in amounts toxic to livestock, so care must be taken in their application.

H. Irrigation

From 50 to 75 cm of water are required to produce a good crop of soybeans. The rate of water use by soybeans can be as much as 5 cm/week. For each 2.5 cm of water available from July 1 to September 20, approximately 134 kg/ha (2 bushel/acre) of soybeans were produced in Illinois (Peters and

Johnson, 1960). In Kansas, 65–75% of the water use occurred during flowering and pod filling (Hay and Pope, 1976). Water use depends primarily on the evaporative demand of the atmosphere over the soybeans. Low humidity, bright sunlight, and dry winds increase water use.

The top meter of clay soil will hold about 20 cm of water at field capacity. Soybean roots have been shown to extract most of their water from the top 150 cm of soil (Stone et al., 1976). Therefore soybeans grown on soil with good water-holding capacity can withstand an extended drought in August or September if the moisture supply has been plentiful before the drought. However, the top meter of sandy soils will hold only about 5 cm of water at field capacity and stress will occur much sooner during a drought.

In areas of low rainfall, irrigation of soybeans is practiced. In more humid areas, generally it does not pay to irrigate soybeans. Cartter and Hartwig (1963) reported profits from irrigation in only 2 of 8 years at Stoneville, Mississippi. Soybeans, unlike corn, do not have a single period during their growth when they are highly susceptible to moisture stress. Soybean yield is largely determined by pod number and the plant can compensate many ways for poor development during a stress period. For example, soybeans flower over a long period and, if early flowers abort, later flowers can make up for the loss. If drought during flowering decreases pod number, seeds per pod and seed size can increase if good conditions return. In Iowa, stress during early flowering reduced yields less than 10%. However, stresses applied for 1-week periods between 4 and 7 weeks after first flowering decreased yields 35–45%. At these later stages of seed development the plant was no longer able to compensate for decreases in all the yield components (Shaw and Laing, 1966). Similar results from other experiments confirm that yields are reduced more by drought during the pod-filling period than at any other time. Therefore, if limited water is available, this is the most profitable period to apply it. Often one irrigation during seed development will cause yield increases as large as several irrigations during the season.

Irrigation produces highest yields on soils with good drainage. Excess irrigation or rainfall can cause waterlogging and decrease yields because of poor root aeration. Excess moisture also reduces nitrogen fixation and leaches soluble nutrients.

IV. HARVESTING

At maturity, complete leaf drop indicates that harvesting can begin within the next week, if good weather prevails. Iowa studies showed that seed moisture could decrease to 15% within 5 days after complete leaf drop (Quick and Buchele, 1974). Harvest losses increase with time after maturity and as seed moisture content decreases. The decision on when to harvest must balance these factors against the need for drying beans if the moisture content is above 13%. Soybeans have been harvested successfully at moisture contents as high as 20%, but seed is often damaged. Little damage occurs at 15%, but general recommendations are to harvest as close to 13% moisture as possible. To achieve this goal growers usually start harvesting at moisture levels above 13% and finish at lower levels. When beans are below 13% moisture, splitting of beans and shattering of pods increase. Harvesting at night or with dew on the plants is a common practice, both to decrease harvest losses and to try to avoid over-dry grain.

Harvest losses with a combine have averaged 9% of the total crop in the United States since 1925 (Quick, 1972). With the widespread, recent use of flexible, floating cutterbars and automatic header height controls, combine losses appear to be decreasing. Without these modifications, Iowa trials showed that about 85% of the combine loss was at the header and about 68% of the loss resulted from effects of the reciprocating cutterbar (Quick and Buchele, 1974). In the Iowa trials, header losses in standing soybeans occurred primarily because stems and attached pods were not collected after they had been cut off. In contrast, Illinois studies showed higher losses due to shattering than stalk losses (Nave *et al.*, 1973). Farm trials in Arkansas, Mississippi, and Illinois indicated both sources of harvest loss were about equal (Nave *et al.*, 1973).

A number of practices can reduce harvest losses. Five seasons of field testing by Quick and Buchele (1974) in Iowa indicated that narrow rows both increased yields and decreased harvest losses because bottom pods were higher off the ground. Illinois results (Nave *et al.*, 1973) also indicate that narrow rows decrease harvest losses unless the plant population is as low as 250,000 plants/ha. At low populations and in narrow rows, soybean plants can branch profusely from lower nodes, increasing the proportion of pods near the ground at harvest time. Higher planting rates raise lower pods and decrease harvest losses. Leaving ridges near the row in cultivation increases harvest losses because the header height must be higher to avoid the soil.

Header loss increases rapidly as bean moisture content drops, mainly because more pods shatter before entering the combine. Quick and Buchele (1974) measured losses of about 4% at 13% moisture and 10% at 10% moisture. Also in their trials, two combines with pickup reels and header height controls were compared after equipping one with a flexible floating cutterbar and one with a standard rigid cutterbar. Average total header loss was 40% lower with the flexible floating cutterbar in one trial and 46% lower in another. Most new soybean combines are equipped with flexible floating cutterbars and header height control. The flexible cutterbars can be added to standard combines and research indicates they will pay for themselves in 50 hours of harvesting or less. The flexible cutterbar is able to follow ground contours and therefore can cut consistently lower. Even with the flexible cutterbar harvest losses are increased by irregular ground surfaces. Soybean land should be levelled before planting, both to facilitate harvesting and to remove wet spots where root rots can develop.

Increasing the forward speed of the combine above 5 km/hour (3 MPH) gives an almost linear increase in harvest losses. Above 7 to 8 km/hour whole plants could slip under the cutterbar without being cut (Quick and Buchele, 1974). Knives with more and narrower sections decrease both losses under the knife and shattering losses, permitting higher combine speeds (Fig. 3). Ideal reel settings for a standing crop are about 15–30 cm above and ahead of the cutterbar with the reel tips travelling 1.25 to 1.5 times faster than the ground speed. Most new combines have hydraulic reel position controls and variable speed reel drives. Pick-up reels and lifters in front of the cutterbar can reduce header losses substantially in lodged soybeans.

Cylinder and separation losses are relatively small. In weed-free conditions they should be about 0.2% of yield (Nave *et al.*, 1973; Byg, 1974). Excessive speed or green weeds can raise threshing losses to 5% of yield. Cylinder speed should be just fast enough to shell the beans. Greater cylinder speeds are necessary at higher moisture levels. Excessive speed causes seed coat damage, with large-seeded varieties being most susceptible, and damage increases at lower seed moisture. Low cylinder speeds are vitally important when harvesting soybeans for seed because seed coat damage greatly reduces germination.

Separation loss results mainly from beans being blown out with the chaff. Adjustment of the fan and sieves should provide clean beans without blowing soybeans out the rear of the combine, or into the tailings, where excessive seed coat damage occurs.

Reducing harvest losses pays big dividends. In a crop yielding 2220 kg/ha (33 bushels/acre), Quick (1973) calculated that a 9% harvest loss represented 33% of profits. To improve his performance, a combine operator should identify where he is losing beans. This is generally done by counting beans on the ground behind the combine, across the total header width. One bean per square meter represents 1.50 kg/ha loss (4 beans per square foot – 1 bushel/acre).

The rapid adoption of better designs in harvesting equipment appears to be reducing losses (Siemens and Hirning, 1974). Further improvements in header design are being tried, including air blasts back over the cutterbar (Nave *et al.*, 1973) (Fig. 4). Row crop headers have performed well (Quick and Buchele, 1974) but decrease the grower's flexibility in row widths. Cir-

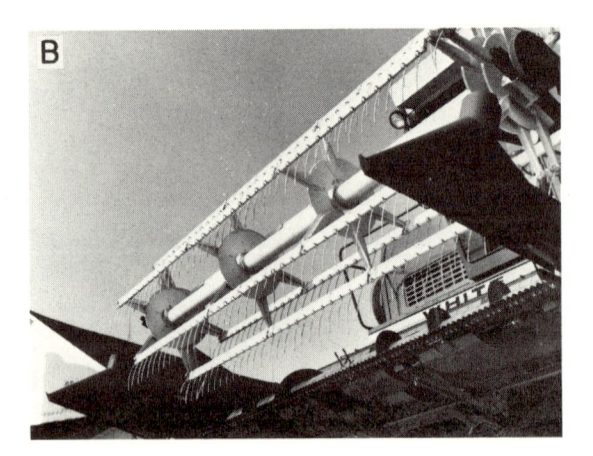

Fig. 3. A. A modern soybean combine harvesting mature soybeans. This combine is equipped with a flexible, floating cutterbar. Note the short stubble height. Photograph courtesy of Deere & Company. B. A closer view of a flexible, floating cutterbar and pick-up reel. The flexible cutterbar floats on the skid shoes seen underneath the platform. This header also has narrow guards and knife sections specially designed to cut more rapidly, reduce cutterbar losses, and permit higher combine speeds. Photograph courtesy of White Motor Corporation.

6. Management and Production

Fig. 4. A. Experimental air-jet guards mounted on a floating cutterbar. Jets of air blow soybeans back onto the combine platform to reduce harvesting losses. B. A side view of the air-jet guards. This combine is also equipped with sensors for automatic header height control. One is shown above the skid plate. Photographs courtesy of W. R. Nave, University of Illinois.

cular cutters rather than knife sections should reduce shattering and knife overload. Other new developments in harvesting soybeans are discussed by Byg (1976). The rise in importance of soybeans is resulting in specialized equipment for this crop rather than machinery designed for cereal grains.

The quality of harvested seed depends greatly upon weather conditions during ripening and harvest. Cool, dry conditions favor good seed quality.

Warm, moist weather increases seed-borne diseases, causes a darkened, weathered appearance of the seed, and frequently reduces germination. Although seed quality usually can be assessed visually, soybeans for seed should be tested for germination.

V. DRYING AND STORAGE

Soybean harvest losses can be reduced by harvesting at moisture levels too high for safe storage. However, beans then have to be dried. Safe moisture levels for storage depend on air temperature, volume stored, and whether beans are to be used for seed. Soybeans can be safely stored for long periods at 11% moisture. The length of time most soybean seed can be stored without losing viability is given by the formula below (Hukill, 1963; Boakye-Boateng and Hume, 1975).

$$\text{Age index} = (\text{storage time in months}) \times 10^{x+y}$$

where $x = 0.143$ times seed percentage moisture and $y = 0.0645$ times storage temperature in °C. In this formula, soybean germination begins to decrease after an age index of 15,000 is reached. This equation says that moisture affects deterioration of seed more than temperature. As an example of use of the equation, seed stored at 11% moisture and 20°C will start to decline in germination percentage after 20 weeks but if the storage temperature is 10°C, seeds will retain a high germination percentage for 90 weeks. These results illustrate why seed can be stored at 14–15% moisture over winter, but when temperatures rise in the spring, deterioration begins. Holman and Carter (1952) found that seed at 10% moisture could be safely stored for 1 year, with only a slight decrease in viability. Seed at 12–12.5% maintained its grade for commercial sale for nearly 3 years, but germination percentage began to decrease during the first year. At 13–14% moisture, seed could be stored during the winter but the germination percentage started to decline during the following spring. Seed for marketing started to deteriorate in the second year. At 14–15% moisture, seed overwintered but started to deteriorate in the spring. The effect of seed moisture on storage is complicated by the fact that moisture migrates with air movement in a bin. Warmer air rises and carries moisture with it, so that the center top of a bin or a peak in a pile accumulates moisture. Holman and Carter (1952) found that farm bins filled with beans at 12 and 13% moisture reached moisture contents of 16–17% near to top center area and the beans there deteriorated. Similar migrations in moisture have been observed in lake freighters storing beans over the winter (D. J. Hume, unpublished data), with subsequent heating and spoilage in the surface layers.

Safe moisture content of beans going into storage, then, depends on how long they will be there and their ultimate use. Soybeans can be dried with conventional grain driers but percentage of cracked beans increases rapidly with drying temperatures. Research at Kentucky (Hamilton et al., 1973) indicated that cultivars differed in susceptibility to cracking. If the drying air was at 21°C, there was little cracking. At air temperatures of 38°, 10–60% of the beans had skin cracks and 5–20% had cracks below the skin. At drying air temperatures of 55°, 50–90% had skin cracks and 20–70% were cracked below the skin.

Low relative humidity in the drying air also causes cracking because drying occurs too rapidly. Air temperatures under 38°C (100°F) and relative humidities of 40–50% are considered satisfactory for most cultivars and for seed.

Drying in the storage bin is most economical for many growers. Bins equipped with a perforated floor are ideal but suitably designed duct systems also will work. Unheated air can be used if it can remove enough water from the beans. Generally this means temperatures above 15°C and relative humidities below 70%. Heat can be added to the airflow using electric resistance heating elements, heat from the motor driving the fan, or other sources. Resistance to air flow in beans is approximately 25% less than for shelled corn, so soybeans can be dried at deeper depths than corn but this is not advised because of the lower temperatures required for drying soybeans.

Soybeans dried in batch or continuous flow driers should be cooled after placing them in a storage bin. This can be done with an aeration system. Ideally, an aeration system should move air down through the grain so the dry outside air passes over the upper, moistest grain. However, drying fans can be used to aerate grain by forcing air up through the grain. Soybeans should be aerated until the grain is 5°C above the outside air temperature or until the grain mass reaches 5°–7°C. A properly designed aeration system can make moving heating grain unnecessary. Specifications for aerating and drying systems are available from many state advisory services.

ACKNOWLEDGMENT

Soybean advisory publications were obtained from the following states: Alabama, Arkansas, Delaware, Florida, Georgia, Illinois, Indiana, Iowa, Kansas, Kentucky, Michigan, Minnesota, Mississippi, New York, North Carolina, Ohio, South Carolina and Wisconsin. This material provided background to develop a consensus of the management practices currently recommended in the major soybean areas in North America. The use of such materials is gratefully acknowledged.

REFERENCES

Andersen, R. N. (1976). *In* "World Soybean Research" (L. D. Hill, ed.), pp. 444-452. Interstate Printers & Publ., Danville, Illinois.
Anderson, G. W. (1974). *Res. Rep. Can. Weed Comm. (East. Sect.)* p. 192.
Anonymous. (1974). *Iowa State Univ., Ext. Serv. Insect Picture Sheet Ser.* No. 6.
Athow, K. C. (1973). *In* "Soybeans: Improvement, Production, and Uses" (B. E. Caldwell, ed.), pp. 459-489. Am. Soc. Agron., Madison, Wisconsin.
Boakye-Boateng, K. B., and Hume, D. J. (1975). *Ghana J. Agric. Sci.* **8**, 109-114.
Bray, R. H. (1961). *Better Crops Plant Food* **45**, 18-19 and 25-27.
Brooks, L. (1976). *In* "Soybean Handbook," Publ. C-449. pp. 18-20. Kansas State University, Manhattan.
Burris, J. S., Edje, O. T., and Wahab, A. B. (1973). *Crop Sci.* **13**, 207-210.
Byg, D. M. (1974). *Ohio State Univ., Ext. Bull.* **578**.
Byg, D. M. (1976). *In* "World Soybean Research" (L. D. Hill, ed.), pp. 197-214. Interstate Printers & Publ., Danville, Illinois.
Calland, J. W. (1949). *Soybean Dig.* **9**, 15-18.
Cartter, J. L., and Hartwig, E. E. (1963). *In* "The Soybean" (A. G. Norman, ed.), pp. 161-226. Academic Press, New York.
Clapp, J. G., Jr., and Small, H. G., Jr. (1970). *Agron. J.* **62**, 802-803.
Cooper, R. L. (1971). *Agron. J.* **63**, 490-493.
Cooper, R. L. (1974). *Agron. Abstr.* p. 51.
Cooper, R. L. (1976). *In* "World Soybean Research" (L. D. Hill, ed.), pp. 230-236. Interstate Printers & Publ., Danville, Illinois.
Cooper, R. L. (1977). *Agron. J.* **69**, 89-92.
Criswell, J. G., and Hume, D. J. (1972). *Crop Sci.* **12**, 657-660.
Criswell, J. G., Hume, D. J., and Tanner, J. W. (1976). *Crop Sci.* **16**, 400-404.
Dunleavy, J. W. (1973). *In* "Soybeans: Improvement, Production, and Uses" (B. E. Caldwell, ed.), pp. 505-526. Am. Soc. Agron., Madison, Wisconsin.
Edwards, C. J., Jr., and Hartwig, E. E. (1971). *Agron. J.* **63**, 429-430.
Garcia L., R., and Hanway, J. J. (1976). *Agron. J.* **68**, 653-657.
Good, J. M. (1973). *In* "Soybeans: Improvement, Production, and Uses" (B. E. Caldwell *et al.*, eds.), pp. 527-543. Am. Soc. Agron., Madison, Wisconsin.
Grabe, D. F., and Metzer, R. B. (1969). *Crop Sci.* **9**, 331-333.
Greer, H. A. L., and Anderson, I. C. (1965). *Crop Sci.* **5**, 229-232.
Hamilton, H. E., Loewer, O. J., Jr., and Overhults, D. G. (1973). *Univ. Ky., Ext. Publ.* **AEN-25**.
Hardy, R. W. F., Holsten, R. D., Jackson, E. K., and Burns, R. C. (1968). *Plant Physiol.* **43**, 1185-1207.
Hartwig, E. E. (1973). *In* "Soybeans: Improvement, Production, and Uses" (B. E. Caldwell *et al.*, eds.), pp. 187-210. Am. Soc. Agron., Madison, Wisconsin.
Hay, D. R., and Pope, D. L. (1976). *In* "Soybean Handbook," Publ. C-449, pp. 12-15. Kansas State University, Manhattan.
Hicks, D. R., and Miller, G. R. (1975). *Minn., Univ., Ext. Folder* **314**.
Holman, L. E., and Carter, D. J. (1952). *Ill., Agric. Exp. Stn., Bull.* **553**.
Hukill, W. V. (1963). *Proc. Int. Seed Test. Assoc.* **28**, 871-873.
Hume, D. J., and Criswell, D. J. (1973). *Crop Sci.* **13**, 519-524.
Huxley, P. A., and Summerfield, R. J. (1974). *Plant Sci. Lett.* **3**, 11-17.
Kennedy, B. W., and Tachibana, H. (1973). *In* "Soybeans: Improvement, Production, and Uses" (B. E. Caldwell, ed.), pp. 491-504. Am. Soc. Agron., Madison, Wisconsin.

Kust, C. A., and Struckmeyer, B. E. (1971). *Weed Sci.* **19,** 147-153.
Littlejohns, D. A., and Tanner, J. W. (1976). *Can. J. Plant Sci.* **56,** 371-375.
Major, D. J., Johnson, D. R., and Leudders, V. D. (1975a). *Crop Sci.* **15,** 172-174.
Major, D. J., Johnson, D. R., Tanner, J. W., and Anderson, I. C. (1975b). *Crop Sci.* **15,** 174-179.
Miller, L. B. (1966). *Ill. Res.* **8,** 19.
Nave, W. R., Tate, D. E., Butler, J. L., and Yoerger, R. R.(1973). *ARS Publ.* **NC-7.**
Nelson, W. E., Rahi, G. S., and Reeves, L. Z. (1975). *Agron. J.* **67,** 769-772.
Nelson, W. L. (1976). *In* "World Soybean Research" (L. D. Hill, ed.), pp. 999-1008. Interstate Printers & Publ., Danville, Illinois.
Panizzi, A. R., Correa, B. S., Gazzoni, D. C., de Oliveira, E. B., Newman, G. G., and Turnipseed, S. G. (1977). *Bol. Tec. Cent. Nac. Pesqui. Soja* No. 1.
Parker, M. B., and Harris, H. B. (1977). *Agron. J.* **69,** 551-554.
Peters, D. B., and Johnson, L. C. (1960). *Agron. J.* **52,** 687-689.
Quick, G. R. (1972). Ph.D. Thesis, Iowa State University, Ames.
Quick, G. R. (1973). *Trans. ASAE* **16,** 5-12.
Quick, G. R., and Buchele, W. F.(1974). *Trans. ASAE* **17,** 1123-1129.
Ryder, G. J., and Beuerlein, J. E. (1977). *Soybean News* **28**(3): 5-6.
Searle, L. (1975). *Successful Farm.* **73**(2), 15-17.
Shaw, R. H., and Laing, D. R. (1966). *In* "Plant Environment and Efficient Water Use" (W. H. Pierre *et al.*, eds.), pp. 73-94. Am. Soc. Agron., Madison, Wisconsin.
Siemens, J. C., and Hirning, H. J. (1974). *University Ill., Agric. Ext. Serv., Circ.* **1094.**
Sinclair, J. B., and Shurtleff, M. C., eds. (1975). "Compendium of Soybean Diseases." Am. Phytopathol. Soc., St. Paul, Minnesota.
Stone, L. R., Teare, I. D., Nickell, C. D., and Mayaki, W. C. (1976). *Agron. J.* **68,** 677-680.
Tanner, J. W., and Hume, D. J. (1976). *In* "World Soybean Research" (L. D. Hill, ed.), pp 342-351. Interstate Printers & Publ., Danville, Illinois.
Turnipseed, S. G.(1973). *In* "Soybeans: Improvement, Production, and Uses" (B. E. Caldwell, ed.), pp. 545-472. Am. Soc. Agron., Madison, Wisconsin.
Turnipseed, S. G., and Kogan, M. (1976). *Annu. Rev. Entomol.* **21,** 247-287.
Van Schaik, P. H., and Probst, A. H. (1958). *Agron. J.* **50,** 192-197.
Voss, R. D.(1967). *Iowa Cert. Seed News* **21,** 8.
Wax, L. M. (1973). *In* "Soybeans: Improvement, Production, and Uses" (B. E. Caldwell *et al.*, eds.), pp. 417-457. Am. Soc. Agron., Madison, Wisconsin.
Weber, C. R. (1966). *Agron. J.* **58,** 46-49.
Welch, L. F., Boone, L. V., Chambliss, C. G., Christiansen, A. T., Mulvaney, D. L., Oldham, M. G., and Pendleton, J. W. (1973). *Agron. J.* **65,** 547-550.
Woods, S. J., and Swearingen, M. L. (1977) *Agron. J.* **69,** 239-242.

7

Processing and Utilization

F. T. ORTHOEFER

I.	Introduction	219
II.	Composition of the Seed	221
	A. Proximate Composition	221
	B. Protein Composition	221
	C. Enzymes	223
	D. Lipid Composition	225
	E. Carbohydrate Composition	226
	F. Biologically Active Constituents	227
III.	Processing	228
	A. Oil Extraction	228
	B. Oil Processing	230
	C. Processing of the Meal	233
IV.	Utilization	236
	A. Soybean Oil	236
	B. Protein Products	238
	C. Textured Proteins	241
	D. Modified Soy Proteins	243
	E. Specialty Uses of Soybean Protein	243
V.	Future Projections	244
	References	246

I. INTRODUCTION

The utilization of soybeans began in the Orient where both medicinal and food value was assigned to this legume. Various types of foods were prepared from it including beverages, pastes, curds, and fermented flavorants, some resembling milk, cheese, and meat. Soybeans were believed to be necessary, along with five other grains, for the existence of the Chinese civilization.

Processing of soybeans into oil and meal started in relatively recent times. Demand for the soybean cake by Japan after 1895 for use as a fertilizer resulted in a sudden demand for this product. The defatted cake was the principal product of the oil mill industry. Although processed soybean oil and meal were introduced into Europe in the 1700's, little interest developed until after the Russo–Japanese War (about 1908) primarily because of its inferior quality compared to the native oil and meal products.

Soybeans were introduced to the Eastern United States in the late 1800's. Production spread to the midwest by 1920. Through the early 1930's, the soybean was grown primarily as a pasture and forage crop. By 1947, however, 85% of the crop was harvested for further processing of the seed.

The initial processing of soybeans into oil and meal consisted of simply pressing the oil from the seed. Solvent extraction of the oil then was found to be a more efficient method for oil recovery and was later adopted.

Soybeans in the United States were considered primarily as an abundant source of inexpensive, edible oil. The oil was used in products such as salad dressings and shortenings where the source of the oil could be hidden. Today, soy oil supplies approximately 60% of the edible oils needs of the United States. The defatted soybean meal served initially as a fertilizer and cattle feed. Its nutritional value in swine and poultry feeding was recognized after World War II and has been used extensively. Today, most of the meal is utilized in animals feeds.

Processing of defatted meal into edible protein products has been limited. The utilization of soybean protein products in human foods is expected to increase dramatically, however, because of the population pressure on the food supply. Apparent consumer acceptance of soybean proteins in foods was demonstrated during the food shortage of 1973–1974. Products such as meat

TABLE I

The Proximate Composition of Soybeans and Seed Parts[a]

	Percentage of whole	Protein	Fat	Carbohydrate	Ash
Whole soybeans	—	40	21	34	4.9
Cotyledon	90	43	23	29	5.0
Hull	8	9	1	86	4.3
Hypocotyl	2	41	11	43	4.4

[a] Wolf and Cowan (1971).

analogues, meat extenders, and dairy substitutes met a receptive market. Functional soy proteins having emulsification, aeration, and texture building properties are now being used by various food processors in bakery, dairy, and meat products.

The following is a general review of the processing and utilization of soybeans. An attempt is made to provide basic information on the commercially important constituents of the soybean and on the conversion of those constituents into marketable products. The review is divided into composition, processing, and utilization sections.

II. COMPOSITION OF THE SEED

A. Proximate Composition

The soybean seed consists of proteins, lipids, carbohydrates, and minerals. The proteins and lipids are the principal parts of commercial interest, accounting for approximately 60% of the seed. These reside mainly in the cotyledon as shown in Table I. Variations, caused by both environmental effects and varietal differences, result in protein levels ranging from 30 to 46% and in oil from 12 to 24%.

The bulk of the protein resides in storage sites called protein bodies or aleurone grains. These subcellular structures are 2–20 μm in diameter and are nearly 75% protein. Protein bodies account for 60–70% of the total protein in the seed. The lipids are concentrated in spherosomes dispersed among the protein bodies and are 0.2–0.5 μm in diameter.

B. Protein Composition

The amino acid composition of defatted soy flour is shown in Table II. Based on human requirements, the essential amino acids are equal to or exceed the levels found in egg protein except for the sulfur-containing amino acids. Methionine is the first limiting amino acid. There are little or no significant differences in the amino acid composition among the soybean cultivars. Soy protein has a high lysine content when compared to most other plant proteins and can be used to increase the nutritional value of plant protein combinations.

Most of the soybean proteins are globulins that are insoluble at their isoelectric point but solubilize upon addition of salt. The isoelectric point occurs at pH 4.2–4.6. Extraction of soy flour with water near neutral pH followed by ultracentrifugation resolves the proteins into four major fractions based on their sedimentation coefficients: 2, 7, 11, and 15 S. Each fraction

TABLE II

Amino Acid Composition of Soybean Protein[a]

Essential amino acids	Meal	Nonessential amino acids	Meal
Lysine	6.9	Arginine	8.4
Methionine	1.6	Histidine	2.6
Cystine	1.6	Tyrosine	3.9
Tryptophan	1.3	Serine	5.6
Threonine	4.3	Glutamic acid	21.0
Isoleucine	5.1	Aspartic acid	12.0
Leucine	7.7	Glycine	4.5
Phenylalanine	5.0	Alanine	4.5
Valine	5.4	Proline	6.3
		Ammonia	2.1

[a] Grams amino acid per 16 grams nitrogen. Rackis et al. (1961).

includes all proteins having similar sedimentation rates. The general composition of each is given in Table III.

The 2 S fraction predominates during early seed development, but by 23 days after flowering the ultracentrifuge pattern is similar to that of the mature seed. At least five proteins have been shown to be present in the 2 S fraction, making up about 20% of the seed protein.

The 7 S fraction represents slightly over one-third of the total soluble proteins. The 7 S globulin, about one-half the 7 S fraction, is a glycoprotein consisting of twelve glucosamine and thirty-nine mannose residues per

TABLE III

Approximate Amounts and Components of Ultracentrifuge Fractions of Water-Extractable Soybean Proteins[a]

Fraction	Percentage of total	Components	MW
2 S	22	Trypsin inhibitors	8,000–21,500
		Cytochrome c	12,000
7 S	37	Hemagglutinins	110,000
		Lipoxygenases	102,000
		β-Amylase	61,700
		7 S Globulin	180,000–210,000
11 S	31	11 S Globulin	350,000
15 S	11	—	600,000

[a] Wolf and Cowan (1971, p. 35).

mole. The 7 S globulin consists of nine subunits which dissociate into 2 S and 5 S proteins at low salt concentrations and pH. At high pH, the 7 S globulin will dissociate into a 0.4 S protein. Disulfide-linked polymers will form with the 7 S globulin resulting in insolubilization of the protein. These occur both in the meal and in pure protein preparations. Mercaptoethanol is used to prevent the disulfide polymerization in laboratory preparations.

The 11 S fraction makes up about one-third of the total soy protein and has only one protein component. It is, therefore, the major protein of soybeans with a molecular weight of 350,000. Twelve subunits, both acidic and basic with approximately 22,300–37,200 MW, make up the 11 S protein. The 11 S fraction can be easily prepared by making a concentrated water extract of soybean meal at 25°–40°C, then cooling the extract to near 0°C. The precipitate that forms is 69–88% 11 S fraction.

The 15 S fraction is only one-tenth of the total protein and may be a polymer of other proteins. Molecular weights of a half million or more have been found. The 15 S protein is also precipitated by chilling of water extracts, as with the 11 S protein.

Soybean proteins are denatured by heat, extremes in pH, and by organic solvents and detergents. During heating at 100°C, the protein approaches a minimum solubility, after which the solubility increases with continued heating. High molecular weight aggregates are formed during heating with gels appearing at protein concentrations near 8%. At 8–12% concentration, the gels will break down at 125°C.

Decreasing the pH of soy protein solutions from 3.8 to 2.0 results in the formation of 2–3 S and 7 S fractions. This is dependent upon the ionic strength. Aggregation appears to be more sensitive to ionic strength in the acid pH ranges. The tendency of the 11 S to dissociate into subunits is counteracted by increasing ionic strength. At pH 12, 14.5% protein solutions form gels. The protein mixtures assume a 3 S sedimentation rate.

Pure organic solvents are less effective in denaturing soy proteins than their aqueous solutions. Alcohols denature the protein more effectively as hydrocarbon chain length increases. The 7 S fraction appears to be the most sensitive to alcohol and the 2 S the least sensitive.

C. Enzymes

The enzymes found in soybeans are listed in Table IV. Commercially, the lipoxygenases are considered of major importance. These catalyze the oxidation of lipids, forming fatty acid hydroperoxides. The hydroperoxides undergo scission and dismutation resulting in the development of off-flavors and aromas. The lipoxygenases are specific for lipids containing a *cis, cis,*-1,4-pentadiene system. Only the 13-hydroperoxide is formed with linoleic

TABLE IV

Enzymes Found in Soybeans[a]

Allantoinase	Lipoperoxidase
Amylases	Lipoxygenase
Ascorbicase	Malic dehydrogenase
Chalcone—flavanone isomerase	α-Mannosidase
Coenzyme Q	Peroxidase
Cytochrome c	Phosphatases
Glycosyltransferase	Phosphorylase
Hexokinases	Transaminases
Lactic dehydrogenase	Urease
Lipases	Uricase

[a] Smith and Circle (1972, p. 159).

acid as the substrate. The 9-hydroperoxide isomer is formed as a result of autoxidation. At least four lipoxygenases are present; two of the lipoxygenases are specific for free fatty acids. Crystalline lipoxygenase from soybeans has been reported to contain no prosthetic group and requires no metal activator or coenzyme. Recent evidence indicates a requirement for an iron acivator.

Hydroperoxides, formed as a result of lipoxygenase activity, readily bleach carotenoid pigments through a coupled, free radical mechanism. The bleaching effect of lipoxygenase is utilized in the bleaching of bread dough by incorporation of enzyme active soy flour. Up to 0.5% soy flour, based on weight of the bread flour, is used in white bread. This is the only commercial application of the enzymes in soybeans.

Other lipid related enzymes present are lipoperoxidase and lipases. Lipoperoxidase activity destroys fatty acid hydroperoxides and is similar to cytochrome c in that preformed linoleate hydroperoxide is used to bleach β-carotene. Lipases catalyze the development of free fatty acids. Soybeans contain two separate lipases. These hydrolyze soybean oil at a faster rate than other vegetable oils.

The α- and β-amylases of soybeans show greater activity to highly branched carbohydrates. Immature seeds contain only 34% of the amylase activity of mature seeds. Their function in mature soybeans is unknown since mature soybean seeds reportedly do not contain starch. Starch, however, may be present. Commercial utilization of the amylases is prevented because of difficulty in isolating pure amylases from the soybean.

Both water-soluble and -insoluble proteases are present in soy. Extraction of the proteases require vigorous agitation during extraction to free the proteases from the meal. The proteases exhibit papain-like activity, acting on

7. Processing and Utilization

interior peptide bonds of proteins. None have been found having trypsin-like activity.

Urease activity, which catalyzes the hydrolysis of urea to ammonia and carbon dioxide, varies with cultivar and growth conditions. Its activity also increases with maturation. Urease is readily inactivated by moist heat and is used as a guide to determine the degree of heat treatment given to the various soy products.

D. Lipid Composition

The lipids of soy include both the triglycerides and phospholipids. Minor components that are lipid-soluble include pigments, tocopherols, sterols, and triglyceride-derived products. Lipids have been, traditionally, the most commercially important constituent of soybeans.

The triglyceride portion represents approximately 95% of the hexane extractables from the seed. The fatty acid composition is shown in Table V. The iodine value of the oil varies with cultivar and climatic environment during growth. The average iodine value is 130. Iodine values as low as 103 to as high as 152 have been found.

Soybean oil belongs to the linolenic acid group of oils along with linseed and hempseed oil. The glyceride composition conforms to an even distribution pattern. The triglycerides, then, each contain two unsaturated fatty acids based on the fatty acid composition. Randomization of the fatty acids increases the melting point of soybean oil.

TABLE V

Fatty Acid Composition of Soybean Oil[a,b]

Fatty acid	Percentage
Myristic	0.1
Palmitic	11.0
Palmitoleic	0.1
Stearic	4.0
Oleic	23.4
Linoleic	53.2
Linolenic	7.8
Arachidic	0.3
Behenic	0.1

[a] Saponification value range, 188–195; iodine value range, 125–138.
[b] From Durkee Industrial Foods Group (1970).

Soybean oil has a tendency to develop off-flavors when exposed to air or light. The development of off-flavors is caused by autoxidation and is referred to as "reversion." The flavors have been described as grassy, fishy, or painty. The major precursor is linolenic acid. A large number of compounds develop during reversion. Some that occur are 2-pentylfuran, ethyl vinyl ketone, pentanol, 4-cis-heptanol, 3-cis-hexanol, and diacetyl.

Soybean oil is a drying oil. "Drying" refers to the transformation of an air-exposed, thin film of oil from a liquid to a firm, tough solid. The process involves a polymerization of the oil through condensation reactions. The "drying" ability of soybean oil is less than obtained with other oils such as linseed and has somewhat limited its industrial applications.

The principal phospholipids of soybean are phosphatidylcholine, phosphatidylethanolamine and phosphatidylinositol. The phospholipids make up 1.5–5% of the crude hexane extractables. Although originally discarded as a sludge, the phospholipids have become valuable additives to food and industrial products. Some of the functional properties of phospholipids include emulsification, wetting, dispersing, and antispattering. Commercially, the phospholipids are marketed as "lecithin." Commercial lecithin, a mixture of phospholipids, is distinguished from chemical lecithin, the choline ester.

Phospholipids are insoluble in acetone, unlike the triglycerides. The acetone insolubility index is used to quantitate phospholipids present in an oil mixture. Acetone extraction can also be used to produce oil-free "lecithin."

Crude soybean oil contains various sterols, the aglycone varieties being campestrol, stigmasterol, and β-sitosterol. Free sterols are the major form found. Sterols are present also as esterified glucosides and acylated glucosides. The distribution of the sterols between the different classes do not change from early seed development through maturity. The sterols are acetone-soluble and are usually present in the crude phospholipid fraction from the oil.

Tocopherols exhibit both vitamin E activity and antioxidant activity in crude soybean oil. Those present in oil include Δ-, γ-, and α-tocopherol. The lecithin fraction from the oil contains approximately 0.1% and the vegetable oil distillate contains approximately 2.5–3.0% tocopherols.

E. Carbohydrate Composition

Soybeans contain about one-third carbohydrates which vary with environmental and varietal differences. Both water-soluble and water-insoluble fractions are present. Defatted soybean flakes contain about 11.6% total soluble sugars. The principal sugars consist of 5% sucrose, 1.1% raffinose, and 3.8% stachyose. Glucose, fructose, galactose, rhamnose, arabinose, glucuronic acid, and verbacose, a pentasaccharide, are also present. Raffinose, a trisaccharide, and stachyose, a tetrasaccharide, cause flatulence in humans because of the absence of α-galactosidase enzymes.

The seed coat comprises the major portion of the insoluble carbohydrates. The insoluble polysaccharides consist of galactomannans, acidic polysaccharides, xylan hemicellulose, and cellulose. Lignin probably is also present. The insoluble carbohydrates of the cotyledons are a mixture of acidic polysaccharides and arabinogalactan. The acidic polysaccharides are regarded as belonging to the pectic group of substances.

F. Biologically Active Constituents

Soybeans contain biologically active constituents that are toxic when fed to various animals. Raw soybean meal will cause inhibition of growth, reduce fat absorption, decrease metabolizable energy of the diet, cause enlargement of the pancreas, and stimulate hypersecretion of pancreatic enzymes in chicks, mice, and rats. Toasted soybean meal does not exhibit these adverse nutritional properties. Trypsin inhibitors are the principal antinutritional agents present. Other biologically active agents present are phytic acid, hemagglutinins, saponins, and phenolic constituents.

There are seven to ten proteinase inhibitors in soybeans. Their biological significance in the bean may be as a metabolic defense against insect or bacterial invasion or for the control of protein hydrolysis during storage or germination. Several trypsin inhibitors (TI) have been isolated from soybeans. Two of these, the Kunitz TI and the Bowman–Birk TI, have been better defined than the other inhibitors. Raw soybean meal contains 1.4% Kunitz and 0.16% Bowman–Birk inhibitor. The Bowman–Birk inhibitor exhibits heat stability due to stabilization through disulfide linkages. The weight loss from trypsin inhibitor activity is caused most likely by loss of the amino acids in the enzymes secreted by the hyperactive pancreas. In addition to the competitive inhibition of trypsin, the Kunitz TI inhibits thrombin and the Bowman–Birk TI inhibits chymotrypsin. TI activity of soy is readily inactivated by heating at 100°C for 15 minutes or by atmospheric steaming at 25% moisture for 20 minutes.

Phytic acid, the hexaphorsphoric acid derivative of inositol, exists in soy as a complex calcium–magnesium–potassium complex referred to as phytin. Phytin affects calcium and zinc nutrition through formation of an insoluble, non-nutritionally available complex. Interaction of phytin with proteins also reduces protein solubility. The protein interaction is pH sensitive. At pH 3.0, the interaction occurs through the cationic groups of the protein. Multivalent cation chelates of phytin form at pH 8.5 and the chelate complex binds with the protein.

The hemagglutinins in soy promote clumping of red blood cells *in vitro*. Hemagglutinins are glycoproteins present at about 3.0% in defatted soy flour. There is no evidence for agglutination of red cells on ingestion of hemagglutinins presumably because of its inactivation by pepsin in the stomach.

Soybean saponins, present in the meal at 0.5%, are glycosides of triterprenoid alcohols. Enzyme inhibition by saponins is nonspecific. Saponins are not absorbed upon ingestion.

The phenolic compounds in soy, genistin and daidzin, exhibit estrogenic activity. Because their estrogenic activity is low, it is doubtful sufficient amounts are present in the soybean to elicit an estrogen-like response in feeds. Conversion to more active compounds during processing or fermentation could possibly occur.

III. PROCESSING

A. Oil Extraction

Approximately 95% of the soybean crop is marketed for processing or exported for processing overseas. With a total annual processing capability in the United States of approximately one billion bushels, large storage facilities are required to assure a continuous supply to the processors throughout the processing year (Fig. 1). Maintenance of quality during storage is essential to avoid economic losses.

Prior to storage, the soybeans are cleaned and dried. At 12% moisture, the beans can be stored for 2 years without loss in quality. At 14% moisture, the beans are usually dried before adding to the silos. Incorrect storage conditions result in various physical and chemical changes to the bean, contributing to inferior quality oil and meal. High moisture generally promotes mold-

Fig. 1. Soybean processing plant. Note the large number of storage silos. (Courtesy of A. E. Staley Manufacturing Co.)

7. Processing and Utilization

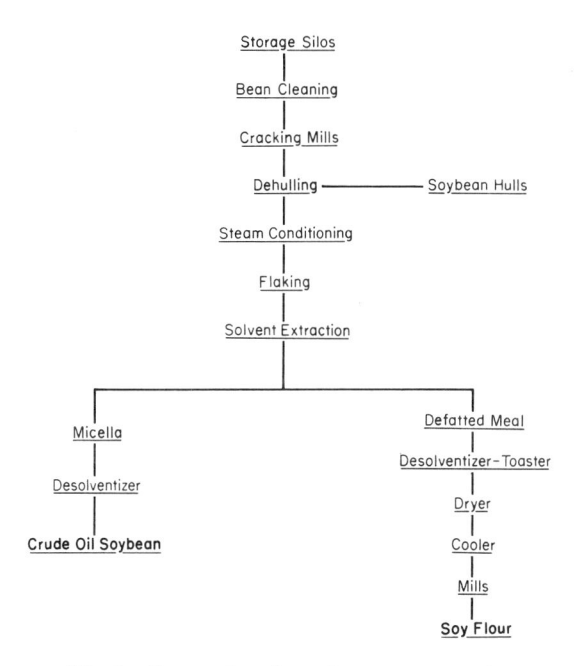

```
                    Storage Silos
                         |
                    Bean Cleaning
                         |
                    Cracking Mills
                         |
                     Dehulling ─────────── Soybean Hulls
                         |
                  Steam Conditioning
                         |
                      Flaking
                         |
                  Solvent Extraction
                  _____|_____
                 |                 |
              Micella          Defatted Meal
                 |                 |
            Desolventizer    Desolventizer-Toaster
                 |                 |
          Crude Oil Soybean      Dryer
                                   |
                                 Cooler
                                   |
                                  Mills
                                   |
                                Soy Flour
```

Fig. 2. Process flow for soybean extraction plant.

ing, high free fatty acid oil, and even charring from heat generation. Aeration is used to prevent moisture condensation. Molding has caused concern over possible development of aflatoxin by *Aspergillus flavus*. Little or no aflatoxin production has been found in moldy soybeans.

The initial products from soybean processing are crude oil and defatted meal or flakes. The processing steps are outlined in Fig. 2. As shown, the beans from storage are cleaned, cracked, dehulled, and flaked prior to oil extraction. Cleaning consists of passing the beans through a magnetic separator to remove iron and steel objects and over various size screens to separate seeds and foreign particles. The beans are cracked by rollers having spiral cut corrugated cylinders. Two to three pairs of rolls crack the beans into six to eight pieces and loosen the hulls. Hull removal by aspiration is necessary to improve extraction efficiency and to decrease the fiber content of the defatted meal intended for edible or poultry feeds. For high fiber feeds, the hulls are added back to the meal after extraction to obtain a 44% protein meal. The defatted, dehulled meal contains 49–50% protein.

The cracked beans are conditioned to about 11% moisture at 63°–74°C, then flaked by passing through differential rolls set at 0.25–0.40 mm clearance. Flaking ruptures the cells of the bean improving solvent contact and overall oil extraction efficiency. Hexane (B.P. 66°–69°C) is generally used for extraction of oil from the flakes. Other solvents and solvent combinations have been proposed to improve extraction efficiency and decrease the flavor of the defatted meal intended for edible uses. Several types of extractors are manufactured although almost all utilize countercurrent extraction. The flakes are conveyed to the extractor and loaded into the extraction baskets. The rate of extraction and solvent temperature influence the residual oil content of the meal. Less than 1% residual oil is desired. Generally, the hexane is kept near 50°C during contact with the flakes.

After extraction, the solvent plus solubilized crude oil, referred to as micella, is first filtered to remove the suspended fines, and the solvent is stripped from the oil by a combination of thin film evaporators and stripping columns. Steam is injected during the final stages of stripping to ensure complete removal of the solvent. The solvent vapors are condensed for reuse and the solvent stripped crude oil is pumped to storage tanks for further processing. Crude soybean oil, because of the presence of natural antioxidants, principally the tocopherols, is storage stable.

The hexane laden flakes are passed through a desolventizer toaster which recovers the solvent and toasts the flakes to obtain optimum nutritional and functional characteristics. The desolventizer toaster consists of a series of steam jacketed compartments with revolving center blades which move the flakes through the unit. Steam is injected during desolventizing to aid solvent removal and adjust the moisture content to 13–15%. The temperature is slowly adjusted to as high as 110°C. After drying and cooling, the flakes are ground into meal.

B. Oil Processing

Crude soybean oil contains from 1–3% of hydratable compounds, primarily the phospholipids (lecithin). Upon addition of warm water to the crude oil the phospholipids form a dense, gum-like, hydrated mass. Commercially, about 1% water is added to the crude oil at 71°C. The hydrated lecithin "gums" are continuously separated from the oil using disk-type centrifuges. The wet gums which contain about 25% moisture are vacuum dried at 100°–110°C. The dried gums are sold as commercial lecithin. Commercial lecithin contains approximately 30% oil and 70% phospholipids as determined by the acetone insolubility method. The dark colored lecithin may be bleached by addition of hydrogen peroxide prior to drying. Partial hydrolyzates are prepared to increase the water dispersibility of the lecithin. Granular, oil-free lecithins are produced by extraction of the entrained oil using acetone. Additional products have been manufactured based on alcohol solubility differences.

Degummed soybean oil contains about 0.02% phosphorus, 0.75% free fatty acids, 1.5% unsaponifiables, and about 0.3% moisture and volatiles. A

7. Processing and Utilization

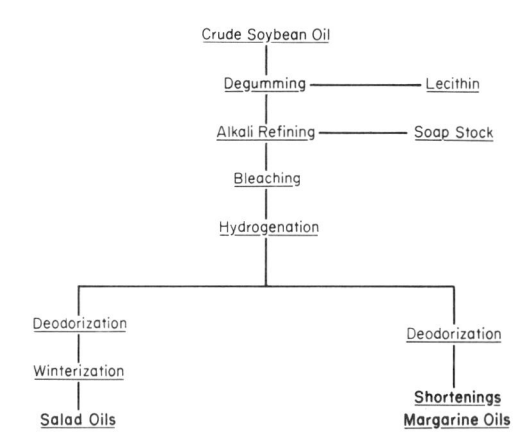

```
                    Crude Soybean Oil
                           |
                     Degumming ————————— Lecithin
                           |
                    Alkali Refining ——————— Soap Stock
                           |
                       Bleaching
                           |
                     Hydrogenation
                    _____|_____
                   |                       |
             Deodorization           Deodorization
                   |                       |
             Winterization                 |
                   |                   Shortenings
              Salad Oils              Margarine Oils
```

Fig. 3. Process flow for production of edible soybean oil products.

clear, bland-flavored oil is required for use in edible products. The steps for the conversion to edible oils as shown in Fig. 3 consist of refining, bleaching, and deodorization. Further processing consists of hydrogenation to impart desired physical properties and/or winterization to remove high melting glycerides from liquid oils.

Removal of the free fatty acids, refining, is the initial step in processing the oil. Alkali is added to the oil for conversion of the free fatty acids to water-soluble soaps. The fatty acid soaps plus excess alkali are then removed by water washes. The alkalis used are sodium hydroxide, sodium carbonate, or a combination. The alkali is added to 60°–70°C oil, mixed, and the soaps removed by centrifugation. Excess alkali is then removed by additional water washes, and the oil is vacuum dried. The free fatty acids in the oil are reduced to 0.05% or less. The recovered fatty acid soaps are referred to as soapstock. Soapstock may be dried as such or acidulated prior to drying. The soapstock, as a byproduct, is used for feed and industrial purposes. The bleaching step after refining improves both the color and flavor stability of the oil. Natural bleaching earths such as Fuller's earth or bentonite and activated carbon is mixed with the oil at 105°–110°C, followed by removal in filter presses. Less than 1% bleaching earth is required. Carbon black is normally used for green oils containing chlorophyll extracted from frost damaged or immature soybeans. The color of the oil is determined using a Lovibond Tintometer which is a graded series of yellow and red glass disks. The color is reported as a given red and yellow value.

The volatiles are removed after bleaching by heating the oil to about 260°C under high vacuum. The deodorization uses vacuums as low as 1–3 mm mercury. Air entering the oil during deodorization results in high color and poor stability and must be minimized. Steam injection is used to speed the process. Deodorization is a continuous process where the oil is allowed to flow over shallow trays within an enclosed chamber. During the initial stages, the temperature is increased to the maximum required. Incoming oil is used to cool the deodorized oil in the final trays. During the latter stages of deodorization, citric or phosphoric acid is added to the oil at about 0.005%. These chelate the traces of heavy metals present which may act as prooxidants. Antioxidants other than metal chelators also are usually added in the final deodorizer tray.

The condensed volatiles obtained from the oil during deodorization contain tocopherols, sitosterol, and stigmasterol and are a primary source for vitamin E and sterol-derived products. These vegetable oil distillates are purchased by the pharmaceutical industry for conversion into marketable products.

After deodorization of the oil, it is filtered prior to storage. Nitrogen gas is used to blanket the oil to prevent oxidation. Antioxidants such as butylated hydroxyl toluene (BHT), butylated hydroxyanisole (BHA), and tertiary butylhydroquinone (TBHQ) may be added to a maximum combined level of 0.02% to prolong the shelf-life of the oil.

Most or the bulk of the soybean oil is sold as refined, bleached and deodorized oil. However, for specific functional and stability properties soybean oil must be hydrogenated. Hydrogenation is usually performed just prior to the deodorization step.

Partial saturation of the highly unsaturated fatty acids results in oxidatively stable soybean oil whereas greater degrees of hydrogenation is used to obtain semi-soft, plastic shortenings. The stability of an oil is usually characterized as the number of hours required to reach a given degree of peroxide equivalents. The active oxygen method (AOM) consists of holding the oil at 100°C with constant bubbling of air and determining the number of hours required to reach 100 meq peroxide value. For example, soybean oil has an AOM of about 10 hours compared to a lightly hydrogenated oil AOM of 25 hours.

The functional properties of hydrogenated oils are characterized by the ratio of solid fat to liquid oil. The solid, high melting glycerides present in hydrogenated oil, is determined by dilatometry. The percentage of high melting glycerides or solids in a fat or oil over the temperature range of 10°–44°C is referred to as the solids fraction index (SFI). The SFI can be manipulated by hydrogenation conditions, blending of oils, and randomization of the fatty acids within the glycerides of various oils.

Hydrogenation most usually is performed with a nickel catalyst. The new copper chromite catalysts are being used to hydrogenate the most highly unsaturated fatty acids more selectively for production of high stability liquid oils. The addition of hydrogen to the double bond of unsaturated fatty acids during hydrogenation is a complex reaction. Upon contact of the fatty acid with the catalyst, hydrogen may add to the double bond, or a transformation from the native *"cis"* to the *"trans"* form may occur without hydrogen addition. The position of the double bond also may move within the fatty acid chain. Therefore, in addition to saturation, positional and geometrical isomers occur. Each isomer plus degree of saturation affects the melting properties of the oil and, subsequently, its functional characteristics.

The degree of selectivity during hydrogenation or the preference for hydrogenation of the most highly unsaturated fatty acids followed by the next most highly unsaturated fatty acid is determined by choice of catalyst and hydrogenation conditions. Highly selective catalysts saturate preferentially the most highly unsaturated fatty acids. For instance, to obtain high stability salad oils, it is desirable to partially saturate only the linolenic acid. Hydrogenation resulting in a high level of positional or geometrical isomers, because of their high melting points, is undesirable since salad oils must remain clear and fluid at refrigerator temperatures.

Batch and continuous hydrogenators are used. Batch converters range from 40,000- to 60,000-pound capacity with approximately a 4-hour cycle. Batch hydrogenators exhibit greater flexibility in hydrogenation conditions whereas continuous units are more suited to large quantity production of a single product type. The temperatures used during hydrogenation vary from 125° to 200°C with hydrogen pressures from 0.5 to 3 atm. Higher temperatures or catalyst concentration generally increases the selectivity. Rapid agitation or starving the catalyst by low hydrogen pressures decreases the selectivity.

Because complete selectivity is generally not obtained in the commercial hydrogenation of soybean oil, some high melting triglycerides are formed. These cause a cloudy haze during cooling of the oil to refrigerator temperature. Commercially, the crystals are filtered from the oil during "winterization." Winterization involves cooling the oil slowly to 13°C over a 12-hour period followed by additional cooling to about 5°C for 18 hours. After holding for 12 hours, the oil is filtered. The filtered oil will remain clear for 20 hours at 0°C.

C. Processing of the Meal

The principal types of defatted meal products based on protein content are shown in the tabulation below.

Product	Percentage protein ($N_2 \times 6.25$)
Meal	44
Flour and Grits	49
Concentrates	70
Isolates	90

The 44% meal is prepared by adding back the soy hulls to the defatted flakes after extraction of the oil. Primary use of the meal is for animal feeds.

The soy flour and grits differ according to their particle size (see tabulation).

Product	Mesh size	Opening (mm)
Grits		
Coarse	10–20	2.00–0.84
Medium	20–40	0.84–0.42
Fine	40–80	0.42–0.177
Flour	100–200	0.149–0.074

In addition to particle size, both flour and grits are manufactured with varying degrees of heat treatment. The treatment of the flakes after leaving the extractor generally determines the properties of the final products and their ultimate use.

Time, temperature, and moisture during desolventizing are the primary variables affecting the product characteristics, principally the protein solubility. Moist heat results in insolubilization of the proteins. The extent of protein insolubilization is determined by various methods although the protein dispersibility index (PDI) or nitrogen solubility index (NSI) are most generally used (Table VI). Examples of various protein solubility ranges of commercially available products are given in Table VII. The NSI of the various products usually range from a low of 7–10 to a high of 90%.

Partially refatted soy flours are prepared for special bakery applications. Low fat flours have up to 6.5% soybean oil added back to defatted flour or may be prepared by blending defatted meal with full fat flour. Full fat flours are processed by steaming whole soybeans to inactivate enzymes, drying to less than 5% moisture, cracking, dehulling, and grinding or extrusion cooking. Lecithinated flours have up to 15% lecithin added to defatted flours.

The grassy, beany, bitter flavor of soy flour results at least partially from oxidation of lipids and from the presence of phenolic compounds. A tremendous amount of effort by many investigators has been made to remove these flavors. The intensity of the off-flavors can be decreased by alcohol extraction or small additions of acetic or other acids. Specialty flours having low flavor profiles are being marketed. The competitive pricing policies of soy flour has limited the extent of the pretreatment that can be made to soy flour. Conversion of flour to higher protein products with better profit margins is economically justified.

Protein concentrates, having 70% protein, were originally produced for nonfood applications such as substitution for casein in paper coatings. Con-

TABLE VI

Terminology of Solubilities and Heat Treatment of Soy Flours[a,b]

Term	Abbreviation	Calculation
% Water soluble nitrogen[c]	WSN	$\dfrac{\text{ml alkali} \times N \times 0.014 \times 100}{\text{Wt of sample}}$
% Nitrogen solubility index[c]	NSI	$\dfrac{\% \text{ WSN} \times 100}{\% \text{ Total nitrogen in sample}}$
% Water soluble protein[c]	WSP	$\% \text{ WSN} \times 6.25$
% Protein solubility index[c]	PSI	$\dfrac{\% \text{ WSP} \times 100}{\% \text{ Total nitrogen in sample} \times 6.25}$
% Water dispersible protein[d]	WDP	$\dfrac{\text{ml alkali} \times N \times 0.014 \times 100 \times 6.25}{\text{Wt of sample}}$
% Protein dispersibility index[d]	PDI	$\dfrac{\% \text{ WDP} \times 100}{\% \text{ Total nitrogen in sample} \times 6.25}$

[a] Wolf and Cowan (1971, p. 39).
[b] Calculations are based on Kjeldahl analysis of extracts where N = normality of alkali, 0.014 = milliequivalent weight of nitrogen, and 6.25 = nitrogen to protein conversion factor.
[c] Based on AOCS Method Ba 11-65. American Oil Chemists' Society (1975).
[d] Based on AOCS Method Ba 10-65. American Oil Chemists' Society (1975).

TABLE VII

Commercially Available Soy Flour and Grits Based on Nitrogen Solubility Index (NSI)[a]

	NSI Range		
Product	20–40	60–70	85+
Grits			
Coarse	X	X	—
Medium	X	X	—
Fine	X	X	—
Flour	X	X	X

[a] From Johnson (1976).

centrates are prepared by removal of the low molecular weight components, mainly the soluble sugars, from soy flour. The remaining carbohydrates consist of an acidic polysaccharide, arabinogalactan, and cellulose. Processing involves immobilization of the proteins and leaching of the soluble components. Immobilization and leaching can be accomplished by: (1) aqueous alcohol extraction, (2) isoelectric extraction, and (3) aqueous extraction of heat denatured flour. Extraction of the soluble carbohydrates from whole soybeans prior to oil extraction is also used.

The composition of the various concentrates is very similar regardless of the extraction method used. Variations in the degree of protein denaturation or insolubilization depends upon the process and the heat treatment. Isoelectric extraction at pH 4.5 causes the least protein insolubilization. The NSI of concentrates, made by the aqueous–alcohol extraction technique, can be increased by various solubilizing treatments. Soy protein concentrates are marketed in particle sizes similar to the flour and grit products.

Soy protein isolates are the purest form of soy proteins available and contain 90% protein. Isolates were initially produced as a milk casein replacer for industrial uses. Isolates are prepared by aqueous or mildly alkaline extraction of defatted flakes or flour having a high protein solubility. The pH of the extraction is usually in the 7 to 9 range. Extraction above pH 9 leads to hydrolytic and rheological alterations of the protein. The extract is separated from the insoluble residue and the clear extract adjusted to pH 4.5 using sulfuric, hydrochloric, or phosphoric acids to precipitate the protein. The insoluble protein is washed, neutralized to pH 6.5–7.0, and spray dried. NSI of isolates can be as high as 90–95%.

Other extraction methods used for soy protein isolate production are isoelectric pH extraction using salt to adjust the ionic strength to solubilize the protein and neutral extraction of flour with water, followed by isoelectric precipitation. The yield of the protein from the meal is highest using alkaline extraction techniques. Yields of 65% or more of the protein are obtained.

IV. UTILIZATION

A. Soybean Oil

The bulk of the soybean oil produced is consumed as a salad oil. Slightly hydrogenated oil is preferred to meet longer shelf-life requirements. Selective hydrogenation is used to limit the formation of saturated, high melting point triglycerides. The iodine value is decreased to approximately 105–110. The hydrogenated salad oils require winterization. Cooking oils from soybean oil are similar to salad oil although a slightly greater degree of satura-

7. Processing and Utilization

tion is used. Iodine values are usually in the 90–100 range. In both salad and cooking oils, the linolenate content is decreased to 3% or less. Development of solids in cooking oils are not as undesirable as in salad oils. In some instances, nearly plastic cooking oils are prepared, supposedly to give a less "greasy" appearance to the finished product.

Shortenings for bakery applications require wide-melting point, plastic fats. To obtain the desired melting and plasticity characteristics, soybean oil is combined with palmitic acid-containing fats, such as cottonseed, lard, or palm oil. The plasticity plus high temperature stability are also dependent upon selectivity during hydrogenation and the ratio of the blends. Fully hydrogenated soybean oil stearine, 0–5 iodine value, may be added to increase the high temperature stability. The solids at 10°C for a shortening would be in the 40% range and at 44°C in the 5–15% range. Higher solids at the high temperature range are desired for fermentation stability and for initial high temperature stability. The high temperature solids also aid in the development of the air cell structure of the baked product. Low temperature plasticity of the shortening plus a crystalline form are needed for ease of creaming into the dough.

Liquid bakery shortenings are manufactured by slowly crystallizing 4–20% of high-melting point glycerides plus emulsifiers and dough conditioners in a liquid oil. Refined and partially hydrogenated soybean oil are used for the liquid oil fraction. The short shelf-life of many bakery products allows the use of the less stable oils. Ease of shipping, storage, and use are contributing to the market acceptance of these industrial products for the baker.

Margarine oils consist of a blend of hard and soft oil fractions. The hard fraction is hydrogenated under nonselective conditions to obtain a high solids profile with rapid melting near body temperature. The soft fraction requires more selective hydrogenation. After combination, cooling results in the formation of a continuous crystalline network of the hard fraction.

Both margarines and shortenings are tempered to develop crystalline forms having the desired melting characteristics and crystal size. The smaller crystals or β' crystals usually are preferred in shortenings for optimum creaming and air incorporation into a dough. Crystallization of shortenings into the desired crystal form is accomplished with the use of swept surface heat exchangers. Initially, the melted shortening is shock-cooled to about 12°–16°C. A second swept surface unit works the supercooled fat for further development of a fine crystalline network. Tempering is completed by storage for up to 3 days in a 26.6°C room. Margarines are tempered after emulsification of the fat, milk ingredients, salt, and other minor ingredients. The degree of crystallization depends upon the product being made. Conventional stick margarine is continuously formed in units with static mixers, and semi-soft products may be filled similar to shortenings.

TABLE VIII

Solid Fraction Indexes of Various Hydrogenated Soybean Oil Products

Temperature (°C)	Solids content (%)		
	No. 1	No. 2	No. 3
10	25	40	52
21.1	12	23	38
26.7	8.0	16	32
31.1	1.5	3	16
37.8	—	—	7
44	—	—	0

Numerous specialty oils based on soybean oil are prepared for formulation of frozen desserts (mellorines), cookie shortenings, confections, icings, ice cream coatings, whipped toppings, and coffee whiteners. The oil portion for each product is prepared to a specific solids fraction index plus melting point for edibility and functionality. The solids profile is obtained by special hydrogenation conditions, blending of oils, or interesterification of mixtures. Examples of various hydrogenated products based on soybean oil are shown in Table VIII.

The industrial uses of soybean oil includes soap manufacture, paints, resins, and drying oil products. Soybean oil is too highly unsaturated for soaps and requires modification. Soybean oil is generally considered a semi-drying oil and has been used in mixtures with stronger drying and faster gelling oils in making varnishes and bodied oils. Soybean oil reacted with maleic anhydride is used for coatings where the maleation improves both the drying and film forming properties of the alkyd resin. Heat bodied, blown, or sulfurized soybean oil is utilized in varnishes for paper, enamels, inks, and stains. Bodying of the oil refers to the processes of taking the oil through the initial stages of polymerization.

B. Protein Products

Utilization of the defatted meal is generally as a protein nutrient for animal or human foods. The forms of protein products presently being marketed are soybean meal, soy flour, soy protein concentrate, and soy protein isolates. Additionally, textured and modified soy proteins are marketed.

Defatted soybean meal was first prepared in the United States in 1915 by expeller pressing. The meal became an important feed ingredient because of the large annual production and high quality protein present.

7. Processing and Utilization

The composition of defatted soybean meal is shown in the tabulation below.

Composition	Percentage
Protein	44–47
Fat	0.5–1.2
Fiber	5.5–6.5
Ash	5.5–6.0
Calcium	0.3–0.33
Phosphorus	0.62–0.65

Widespread utilization of the meal in poultry rations began when observations on heat inactivation of antinutritional factors in raw beans improved feed efficiency in poultry nutrition. Nutritional investigations showed the most efficient growth rate was obtained using a 20% protein ration for broilers. The advantages of soybean meal for swine was later shown. The use of soybean meal in animal feeds accounts for almost 98% of the processed products.

Human food usage of soybean protein products accounts for only about 2% of the total meal production. Several forms of protein products are available. These generally are classified as: (1) flours and grits, (2) concentrates, and (3) isolates.

Utilization of soy proteins in foods is for functional properties as well as nutritional fortification. The functional properties along with various applications are outlined in Table IX. Functionality of soy is related to the surface active properties of the protein, gelation, and fat and water absorption.

The surface-active properties are derived from the tendency of the protein to orient at the oil–water interface of emulsions resulting in a lowering of the interfacial tension. The protein also increases the viscosity of the emulsion aiding in the prevention of fat droplet coalescence. Generally, the high NSI soy protein products are required, although the relationship of NSI to emulsification power is not absolute. Soy protein emulsification is useful in food products such as soups, gravies, sauces, and meat products.

Soy protein isolates will form a gel when heated to 70°–100°C for 30 minutes. Disulfide bonding is involved in the gelation. The conversion of soy proteins to a gel structure is, unlike gelatin, irreversible under the above conditions unless temperatures of about 125°C are used. The irreversible gelling is especially important for texture building in simulated or extended meat products. Soy protein concentrates and isolates are used to replace nonfat milk solids in comminuted meat products because of the gel-forming character and emulsion stabilizing function.

TABLE IX

Functional Properties of Soybean Proteins[a]

Property	Protein form used[b]	Food system
Emulsification		
Emulsion formation	F,G,C,I	Frankfurters, bologna, sausages
	F	Breads, cakes, soups
	I	Whipped toppings, frozen desserts
Emulsion stabilization	F,G,C,I	Frankfurters, bologna, sausages
	F	Soups
Fat absorption		
Promotion	F,G,C,I	Frankfurters, bologna, sausages, meat patties, simulated meats
Control	F,I	Doughnuts, pancakes
Water absorption		
Promotion	F,C	Breads, cakes, confections, simulated meats
Control	F	Macaroni
Retention	F,C	Breads, cakes, confections
	C	Meat patties
Texture		
Viscosity	F,C,I	Soups, gravies, chili
Gelation	F,C,I	Ground meats
	I	Simulated ground meats
Shred formation	F,I	Simulated meats
Chip and chunk formation	F	Simulated meats, fruits, nuts, and vegetables
Fiber formation	I	Simulated meats
Spongy structure formation	I	Simulated meats, dried tofu
Dough formation	F,C,I	Baked goods
Adhesion	C,I	Sausages, luncheon meats, meat patties, meat loaves and rolls, boned hams
Cohesion	F,I	Baked goods
	F	Macaroni
	I	Simulated meats
	I	Dried tofu
Elasticity	I	Baked goods
	I	Simulated meats
	I	Gels
Film formation	I	Frankfurters, bologna
Color control		
Bleaching	F	Breads
Browning	F	Breads, pancakes, waffles
Aeration	I	Whipped toppings, chiffon mixes, confections

[a] Wolf and Cowan (1971, p. 52).
[b] F,G,C, and I represent flours, grits, concentrates, and isolates, respectively.

The water/fat absorption property is probably the most useful function of soy protein. Soy flour absorbs two times or more its weight of water and concentrates will absorb three to five times. Water absorption is related to the hydrophilic nature of the protein. Water absorption is least near the isoelectric pH. Addition of soy flour to bread dough increases the dough yield, permits addition of more water to the dough, and improves dough handling. Addition to ground meats increases frying yield because of the water holding power. The mechanism of fat absorption is not well known although it may relate to the emulsification properties. Fat binding by soy flour in extended meats decreases cooking losses of the fat and provides a more tender, flavorful product. Conversely, soy flour use in some products limits fat absorption during deep frying, for example, in doughnuts and batter mixes. Apparently an impermeable film of denatured protein is formed between the dough and the oil.

On the basis of analysis, soybean protein is of high nutritional quality. However, not only quality but digestibility, species involved, and presence of inhibitors can affect the nutritional value of soy protein. The superior quality of heated versus raw soybean meal has been noted. Toasting of soybean meal for animal feeds is of less significance than for food uses of soy. Edible soy requires functional properties which are generally lost upon toasting.

The protein efficiency ratio (PER) of soy flour, compared to casein at 2.50, is about 2.20. Fortification of soy flour with methionine, the first limiting amino acid, increases the PER to nearly equivalent to that of casein. The PER is somewhat dependent on degree of toasting since high NSI flours may have a PER as low as 1.80.

Soy protein concentrates, based on analysis, do not differ significantly from flour in amino acid composition. Wide variations in PER between processors of concentrates do occur, most likely because of differences in heat treatment of the flour and in the processing of the concentrate. Soy protein concentrates are available with PER nearly equivalent to toasted flour. Methionine fortification, similarly to flour, can result in a PER equivalent to casein.

The nutritive value of commercial protein isolates is highly variable. During preparation of isolates, fractionation of the protein occurs and, since particular care is taken to ensure protein solubility, proteinase inhibitors may be present. The PER of isolates ranges from approximately 1.1 to 1.8.

C. Textured Proteins

Textured soy proteins refer to modified proteins having meat-like textures. The textured products are used for preparation of pet foods simulating

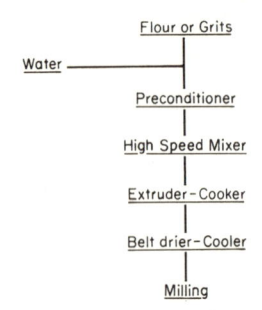

```
                    Flour or Grits
         Water ─────────┤
                    Preconditioner
                        │
                   High Speed Mixer
                        │
                   Extruder-Cooker
                        │
                   Belt drier-Cooler
                        │
                      Milling
```

Fig. 4. Process flow for extrusion of soy protein products (Smith and Circle, p. 315, 1972).

meats, extension of meats, and simulated edible meats. Textured soy flour in the United States school lunch program alone consumes approximately 40 million pounds annually, primarily as a ground meat extender. In addition to textured flours, concentrates and isolates are manufactured having a chewy, resilient texture and appearance for extension of flaked-formed meats, emulsion meats, and ground meat.

Various methods are used to texturize soy products. The primary process is extrusion (Fig. 4). Extrusion texturization involves mixing soy flour, concentrate, or isolate with water, feeding to a continuous cooker-extruder, heating under pressure, and extruding. The heated, compressed mass expands on extrusion resulting in a sponge-like mass. After hydration, the textured product has a chewy, resilient texture similar to meat. Various sizes, flavors, and colors of textured soy proteins are marketed.

Soy proteins can be texturized using a process similar to the spinning of synthetic fibers through a multiholed spinneret die. The protein is solubilized in an alkaline solution and forced through the spinneret die immersed in an acid coagulating bath. Only protein isolates can be used because nearly complete solution of the protein is necessary. The protein immediately coagulates when issuing from the die, giving a continuous protein filament. The filaments are stretched to orient the proteins within the fibers, giving control over the toughness of the filaments. Filament diameters are approximately 75 μm.

The filaments after grinding are used as meat extenders. Groups of filaments, when mixed with binders such as egg white, and suitable flavors and colors, followed by compressing and heat setting, form imitation meats such as chicken, ham, or beef.

Additional methods that are being used to texturize soy proteins include extrusion through spinneret dies avoiding the alkaline solubilization step, high temperature compression, and freeze–thawing of gelled proteins.

These methods are used to a lesser extent than either extrusion or alkaline spinning.

D. Modified Soy Proteins

Modified soy proteins are used as adhesives, coatings, and aeration aids. These are intended to replace or extend animal-derived proteins such as casein, gelatin, or egg albumin. Competition from petroleum derived products has limited the industrial markets for proteins, whether animal or vegetable based, but the recent price increases in petroleum products has brought renewed interest in industrial modified soy proteins.

The soy proteins marketed for paper coatings are hydrolyzed under alkaline conditions, precipitated at pH 4.5, and dried. The alkaline hydrolysis alters the quaternary structure of the protein and hydrolyzes some of the covalent bonds with liberation of ammonia and sulfites. In use, the modified protein is solubilized using ammonium hydroxide. High solids solutions of up to 30% are formulated without gelling of the protein. Specific rheological and adhesive properties are dependent upon the pH, time, and temperature of hydrolysis. Low-, medium-, and high-viscosity grades of industrial soy proteins for paper coatings are manufactured.

Aeration aids for extension of gelatin and egg albumin in confectionery and bakery products are prepared by pepsin hydrolysis of soy flour. These powerful aerators rapidly form foam structures in marshmallows, angelfood cakes, bar mixes, and frozen ices. The pepsin-hydrolyzed proteins do not heat coagulate or gel as do the animal-derived proteins.

E. Specialty Uses of Soybean Protein

Almost 7% of all infants exhibit some degree of allergenicity to cow's milk. Soy milk, although intended for usage as a nutritional replacement for cow's milk, has been an attractive alternative in infant nutrition.

Originally, soy milk was processed from whole soybeans. The beans were allowed to soak in water after a preliminary wash, and then were ground. Additional water was added to adjust the final solids to the desired concentration. The slurry was then heated to near boiling for 15–20 minutes to improve the nutritive and flavor qualities of the milk. The insoluble solids were removed prior to packaging and distribution. Flavors are generally added to improve the acceptability. Process variations in soy milk production have included the use of α-galactosidase and invertase to decrease flatulence. The heat treatment and time within the process apparently controls the development of off-flavors. Lipoxidase inactivation prior to grinding of the beans is necessary.

Attempts to dry soy milk show that upon aging there is a loss in dispersibility. This is most likely caused by protein polymerization through disulfide cross-linkages. Reducing agents such as sodium bisulfite improve the dispersibility.

Special polyunsaturated cow's milk can be obtained by manipulation of the cow's diet. Simple addition of raw soybeans to the cow's diet increases the stearic acid content of the butterfat. If the dietary lipids are protected from hydrogenation in the rumen, the fatty acid distribution in milk fat is changed significantly. Formation of highly cross-linked protein capsules about the fat globules has been used to bypass the rumen. Simple grinding and homogenization of full fat soybeans followed by cooking with formaldehyde gives an effective barrier to hydrogenation of oil and results in transfer of the dietary lipids to the milk fat.

Formulated edible products also use soy proteins as a casein replacer. Particular products such as coffee whiteners, whipped toppings, and frozen desserts are marketed. Protein fortified beverages, cereals, and snack food items have been made. Fortified beverages, however, usually require acid-soluble proteins. Special enzyme hydrolysates or hydrolysate fractions have been found that have acid solubility.

V. FUTURE PROJECTIONS

The development of markets for soybean oil and meal has led to phenomenal growth of the soybean industry in the past 50 years. The growth rate suggests that the crop will continue to expand but most likely somewhat slower than experienced in the past.

In 1974, 57% of the soybeans harvested in the United States were processed into oil and meal and utilized in the United States. The remainder was exported as whole beans, meal, and oil. Most of the oil (93%) is utilized in foods primarily as salad oils and shortenings (Table X). The remainder consists of industrial usage (3%) and foots plus losses (2%). Marketing of soybean oil has become more difficult recently and oil, in fact, has often been in surplus. In the near future, supplies of vegetable oils will probably be in excess of demand because of the increasing availability of palm and lauric oils (coconut and palm kernel). The supply of palm oil is expected to increase dramatically through the 1980's because of maturing of palm plantations. The output of soybean oil will be even more dependent upon the demand for the meal.

In recent years, the demand for meal has been stronger than for the oil. Soybean meal presently supplies approximately 61% of the world's meal consumption. This is an increase from 49% just 10 years ago. The increase in

TABLE X

Domestic Use of Soybean Oil in 1973[a]

Use	Amount (kg × 10^6)
Food	
Shortening	1098
Margarine	769
Cooking and salad oils	1488
Other edible	11
Nonfood	
Paint and varnish	43
Resins and plastics	30
Other drying oil products	1.4
Other inedible	9.5
Foots and losses	134

[a] From U.S. Department of Agriculture (1977).

soybean meal's position occurred because of its favorable price and supply compared to other meals.

The increased overall wealth of the world has been reflected in an increased demand for meats of all kinds. Because of this demand for meats, and the increased demand for soybean meal as an animal feed, it is now the principal product of the soybean milling industry. The future demand for the meal will depend upon general world economy. Most likely, the price of the meal will have to increase significantly to offset the decreased demand for the oil.

TABLE XI

Estimate of Soybean Proteins Produced as Food Ingredients in 1974 and Projections for 1985[a]

Protein product	1974 (kg × 10^6)	1985 (kg × 10^6)
Flours and grits	409	909
Concentrates	31.8	227–273
Isolates	27.3	180–227
Textured products		
Flours and grits	40.9	180–227
Isolates	4.6	

[a] From Wolf (1976).

The direct consumption of soybean protein in human foods accounts for only 26.5 million bushel of the 1.26 billion bushel crop (Table XI). The utilization of soybean proteins as food ingredients is expected to increase three to four times by 1985. By 1980 simulated meats and meat extenders are projected to reach a sales volume of nearly $2 billion. But even by 1985, the expected direct food usage of soybeans will only account for approximately 5% of the crop. The remainder will be converted into meat, milk, and eggs.

The conversion of soybean proteins into animal products is inefficient. The world is faced with shortages of land, energy, water, and fertilizer. The efficiency of the direct consumption of soy proteins is obvious. Many food processors are already replacing traditional proteins in formulated food products with less expensive plant proteins. As more is learned about soybean protein and methods to instill functionality similar to the traditional animal proteins, greater and greater substitutions will be made in food products. Beyond 1985, edible proteins from soybeans are likely to continue to utilize an increasingly larger portion of the crop.

REFERENCES

American Oil Chemists' Society. (1977). "Official and Tentative Methods," 3rd ed. Am. Oil Chem. Soc., Champaign, Illinois

Durkee Industrial Foods Group. (1970). "Composition and Constants of Edible Fats and Oils." SCM Corporation, Cleveland, Ohio.

Hanson, L. P. (1974). "Vegetable Protein Processing." Noyes Data Corporation, Park Ridge, New Jersey.

Johnson, D. W. (1976). *J. Am. Oil. Chem. Soc.* **53**, 321.

Markley, K. S. (1950). "Soybeans and Soybean Products," Vol. I and II, Wiley (Interscience), New York.

Rackis, J. J., Anderson, R. L., Sasame, H. A., Smith, A. K., and Van Etten, C. H. (1961). *J. Agric. Food Chem.* **9**, 409.

Smith, A. K., and Circle, S. J. (1972). "Soybeans: Chemistry and Technology," Vol. 1, Avi Publ. Co., Westport, Connecticut.

Swern, D. (1964). "Bailey's Industrial Oil and Fat Products." Wiley (Interscience), New York.

U.S. Department of Agriculture. (1977). "Fats and Oils Situation," p. 10, FOS 275. Econ. Res. Serv., USDA, Washington, D.C.

Wolf, W. J. (1976). *Adv. Cereal Sci. Technol.* **1**, 331.

Wolf, W. J., and Cowan, J. C. (1971). "Soybeans as a Food Source," p. 13. Chem. Rubber Publ. Co., Cleveland, Ohio.

Subject Index

A

Agronomic characteristics
 effects of light, 78-89
 effects of pests, 95-97, 105-113
 effects of temperature, 88-95
 effects of water, 97-102
 effects of weeds, 113-115
 effects of wind, 102-105
Alachlor, 174
Atrazine, 187

B

Backcrossing, 132-134
Bacterial pustule, 107, 197
Bacteroids, 22
Bentazon, 186
Blends, 120, 152-153
Breeder seed, 130, 131
Breeding, 120-155
 objectives, 120-127
 cultivar development, 127-132
 methods, 132-143
 operations, 143-153
Brown stem rot, 106

C

Carbon assimilation, 46-62
Carbon dioxide diffusion, 47-48
Certified seed, 130-132
Cocklebur, 186
Cultivar development, 127-132

D

Day length, 3, 35-37, 78-89, 161
Defatted cake, 220
Dinoseb, 186
Diseases, 188-199
 bacterial, 196-198
 fungal, 190-196
 viral, 198-199
Double cropping, 172

E

Early generation testing, 138-9
Economics, 12
Exports, 12-14, 244

F

Flooding, 101
Flower induction, 26, 35-36
Flower structure, 24, 27, 143-146
Foundation seed, 130, 131
Fungi, 190-196
 Aspergillus flavus, 96
 Aspergillus glaucus, 96
 Fusarium spp., 95, 185
 Macrophomina phaseolina, 95
 Phakopsora pachyrhiza, 95, 104, 107
 Phialophora gregata, 106
 Phytophthora megasperma, 31, 95, 107, 126
 Pythium spp., 95, 185
 Pythium aphanidermatum, 95
 Pythium debaryanum, 95
 Pythium ultimum, 95
 Xanthomonas phaseoli, 107

G

Glycine max, 1, 17
Glycine ussuriensis, 1
Growth analysis, 28-30, 49-50
Growth regulators, 41

H

Hail injury, 201, 207
Harvesting, 151, 209-214
  equipment, 210-213
  losses, 210-211
Herbicides, 185-187
Hybrid seed, 120, 153-155

I

Insects, 109-113, 201-206
  Colias eurytheme, 109
  Cynthia cardui, 96
  Empoasca fabae, 103
  Epilachna varivestis, 96, 109
  Nezara viridula, 112
  Plathypena scabra, 96, 109
  Sericothrips variabilis, 103
  Spissistilus festinus, 111
  Tetranychus turkestani, 96
Irrigation, 208-209

L

Leaf,
  chlorophyll content, 51
  pubescence, 48
  stomatal frequency, 48
  structure, 25-26
Lectins, 22
Leghemoglobin, 15, 64
Linoleic acid, 94, 126
Linolenic acid, 94, 126, 225-226
Linuron, 174
Lodging, 123, 187-188

M

Management, 157-214
  harvesting, 209-214
  inoculation, 181-185
  pests, 188-206
  planting, 159-185
  weed control, 168-170, 185-187
Maturity groups, 3, 79-80, 161-162
Meal, 12-13, 220, 233-236, 244-245
  processing, 233-236

Methionine, 126, 221
Minor element deficiencies, 207-208
Mosaic, 107-108
Mulching, 91, 92, 98-99

N

Nematodes, 108, 189, 200
  Heterodera glycines, 107, 200
  Meloidogyne hapla, 95, 189
  Meloidogyne incognita, 189
  Rotylenchulus reniformis, 200
Nitrate reduction, 70-72
Nitrogen,
  effect on fixation, 39, 63-64
  effect on nodule formation, 39, 69
Nitrogenase, 66, 67
Nitrogen assimilation, 62-72
Nitrogen fixation, 6, 15, 22, 39, 40, 63, 65-70, 92, 165
  acetylene reduction assay, 65-66
  environmental effects, 67-69, 92-93
  stable isotope method, 65

O

Oil, 13, 220, 225, 228-233
  extraction, 228-230
  processing, 230-233
  utilization, 236-238

P

Paraquat, 174
Pedigree seed, 130, 131
Pedigree selection, 135-136, 143
Pests, 95-97, 105-115, 189-201
Plant development, 30-41
  daylength, 35-37
  light intensity, 35, 50, 52, 87-88
  moisture, 31-33, 99-102
  soil aeration, 37
  soil fertility, 37-41
  temperature, 33-35, 89-91, 161
Planting dates, 84-87, 170-172, 174, 177
Planting depth, 180
Planting equipment, 180-181
Photoperiod, 26, 36, 78-84, 121, 171
Photorespiration, 46, 49
Photosynthate distribution, 61-62
Photosynthesis, 45-62
  environmental effects, 51-55, 92, 99, 161
  internal control mechanisms, 58-61

Subject Index

Phytophthora rot, 95, 107, 126, 127
Pollination, 27, 143-145, 154
Processing, 220, 228-236
Production,
  Brazil, 10-11, 12
  Canada, 5
  United States, 1-9
  Worldwide, 10-12
Pythium rot, 95

Q

Qualitative characters, 121-123, 130
Quantitative characters, 121, 123, 128, 130

R

Recurrent selection, 139-141
Rhizobia in soil, 38-39
Rhizobium japonicum, 20, 22, 38, 63, 64, 91-92, 108, 165, 200
Ribulose diphosphate (RuDP), 49, 54
Root development, 19-20, 99
Root nodules, 20-22, 39, 64-69, 92, 200
Row width, 176-179
Rust, 95, 104

S

Seedbed preparation, 150, 167-168
Seed composition, 46-47, 94, 124, 220-228
Seed development, 28
Seed germination, 31, 33, 37, 89, 97, 100
Seeding rate, 175-180
Seed morphology, 18-19
Seed quality, 93
Seed storage, 93-94, 214-215, 228
Shoot morphology, 22-28
Single-seed descent, 136-138, 141, 143
Soil requirements, 163-167
  lime, 166-167
  nitrogen fertilizer, 164-166
  nutrient element deficiencies, 166-167
Soy enzymes, 223-225
Soy flour, 234
Soy lecithin, 226-230
Soy lipids, 225
Soy milk, 243
Soy protein, 221-223, 234, 236
  amino acids, 222-223
  protein products, 238-244

T

Trifluralin, 186

U

Utilization, 12-14, 220-221, 236-246

W

Weed control, 168-170, 174, 178, 185-187
Weeds, 113-115
  Abutilon theophrasti, 114
  Amaranthus hybridus, 114
  Cassia obtusifolia, 114
  Cyperus rotundus, 114, 186
  Ipomoea hederacea, 114
  Ipomoea purpurea, 114
  Setaria faberii, 113
  Setaria lutescens, 113
  Sorghum halepense, 115
  Xanthium pennsylvaniaum, 114
Wind, 102-105

Y

Yields, 2, 5, 7, 8, 9
Yield testing, 146-150